T0237801

Lecture Notes in Computer Science 4837

Commenced Publication in 1973
Founding and Former Series Editors:
Gerhard Goos, Juris Hartmanis, and Jan van Leeuwen

Editorial Board

Mirosław Kutyłowski
Jacek Cichoń   Przemysław Kubiak (Eds.)

Algorithmic Aspects of Wireless Sensor Networks

Third International Workshop, ALGOSENSORS 2007
Wrocław, Poland, July 14, 2007
Revised Selected Papers

 Springer

Volume Editors

Mirosław Kutyłowski
Jacek Cichoń
Przemysław Kubiak
Wrocław University of Technology
ul. Wybrzeże Wyspiańskiego 27 50-370 Wrocław, Poland
E-mail: {Miroslaw.Kutylowski, Jacek.Cichon, Przemyslaw.Kubiak}@pwr.wroc.pl

Library of Congress Control Number: 2008920058

CR Subject Classification (1998): F.2, C.2, E.1, G.2

LNCS Sublibrary: SL 5 – Computer Communication Networks
and Telecommunications

ISSN   0302-9743
ISBN-10   3-540-77870-5 Springer Berlin Heidelberg New York
ISBN-13   978-3-540-77870-7 Springer Berlin Heidelberg New York

Typesetting: Camera-ready by author, data conversion by Scientific Publishing Services, Chennai, India
Printed on acid-free paper   SPIN: 12222025   06/3180   5 4 3 2 1 0

Preface

Wireless ad-hoc sensor networks have recently become a very active research subject due to their high potential of providing diverse services to numerous important applications, including remote monitoring and tracking in environmental applications and low-maintenance ambient intelligence in everyday life. The effective and efficient realization of such large-scale, complex ad-hoc networking environments requires intensive, coordinated technical research and development efforts, especially in power-aware, scalable, robust, wireless distributed protocols, due to the unusual application requirements and the severe resource constraints of the sensor devices.

On the other hand, a solid foundational background seems necessary for sensor networks to achieve their full potential. It is a challenge for abstract modeling, algorithmic design and analysis to achieve provably efficient, scalable and fault-tolerant realizations of such huge, highly dynamic, complex, nonconventional networks. Various features, including the extremely large number of sensor devices in the network, the severe power, computing and memory limitations, their dense, random deployment and frequent failures, pose new interesting abstract modeling, algorithmic design, analysis and implementation challenges of great practical impact.

This workshop aimed to bring together research contributions related to diverse algorithmic and complexity theoretic aspects of wireless sensor networks. This was the third event in the series. ALGOSENSORS 2004 was held in Turku, Finland, ALGOSENSORS 2006 was held in Venice, Italy. Since its beginning, ALGOSENSORS has been collocated with ICALP.

The Third International Workshop on Algorithmic Aspects of Wireless Sensor Networks (ALGOSENSORS 2007) was organized by Wrocław University of Technology. The workshop was held on July 14, 2007, in conjunction with ICALP 2007. After a careful review by the Program Committee of 25 submissions, 11 regular papers were accepted to ALGOSENSORS 2007. Apart from the regular talks, two keynote speeches were given at the workshop.

October 2007

Mirosław Kutyłowski
Jacek Cichoń
Przemysław Kubiak

Organization

Conference and Program Chair

Mirosław Kutyłowski (Wrocław University of Technology, Poland)

Program Committee

Alan A. Bertossi, University of Bologna, Italy
Costas Busch, Rensselaer Polytechnic Institute, USA
Bogdan Chlebus, University of Colorado at Denver, USA
Jacek Cichoń, Wrocław University of Technology, Poland
Andrea Clementi, University of Rome "Tor Vergata," Italy
Tassos Dimitriou, Athens Information Technology, Greece
Sándor Fekete, Braunschweig University of Technology, Germany
Eric Fleury, INRIA, France
Rachid Guerraoui, EPFL, Switzerland
Dariusz Kowalski, University of Liverpool, UK
Evangelos Kranakis, Carleton University, Canada
Mirosław Kutyłowski, Wrocław University of Technology, Poland
Jan van Leeuwen, Utrecht University, The Netherlands
Sotiris Nikoletseas, University of Patras and CTI, Greece
Pekka Orponen, Helsinki University of Technology TKK, Finland
Jose D.P. Rolim, University of Geneva, Switzerland
Christian Scheideler, TU Munich, Germany
Christian Schindelhauer, University of Freiburg, Germany
Paul G. Spirakis, University of Patras and CTI, Greece
Sébastien Tixeuil, Univ. Paris-Sud / INRIA, France
Peter Widmayer, ETH Zurich, Switzerland

Organizing Committee

Ioannis Chatzigiannakis, Publicity Chair, University of Patras and CTI, Greece
Agnieszka Różańska, Local Organization, Wrocław University of Technology,
    Poland

Steering Committee

Josep Diaz, Technical University of Catalonia, Spain
Jan van Leeuwen, Utrecht University, The Netherlands

Sotiris Nikoletseas (Chair), University of Patras and CTI, Greece
Jose Rolim, University of Geneva, Switzerland
Paul Spirakis, University of Patras and CTI, Greece

Publicity Chair

Ioannis Chatzigiannakis, Publicity Chair, University of Patras and CTI, Greece

External Referees

Emanuele Fusco
Luciano Guala'
Maciej Gębala
Tomasz Jurdziński
Marcin Kik
Alexander Kröller
Jarosław Kutyłowski
Marcin Zawada

Sponsoring Institutions

Polish Ministry of Science and Higher Education

Wrocław University of Technology

Table of Contents

Algorithmic Challenges for Sensor Networks – Foreword to ALGOSENSORS 2007

Mirosław Kutyłowski

Institute of Mathematics and Computer Science,
Wrocław University of Technology, Poland
`miroslaw.kutylowski@pwr.wroc.pl`

Advances in sensor technology lead to new research and technology challenges that are essential for success or failure of sensor networks in practice. Solutions for these challenges have fundamental importance for determining direction and scope of practical applications. At the same time, many brilliant solution ideas from the early stage of development of sensor networks have to be reconsidered and adjusted to the changing state of technology and practical limitations that we were not aware of.

It turns out that design of sensor networks is an interdisciplinary task. Problems arising for sensor networks concern, among others, radio communication issues, hardware design and quite specific distributed algorithms. For instance, limitations of bandwidth in wireless communication and need to preserve energy impose severe demands on intelligent data analysis and compresssion before a sensor actually starts a transmission. So, despite simplicity of a sensor, sophisticated but efficient methods of data analysis must be applied on site. A layered design, so successful for wired communication protocols and operating systems, might be of little use here. Many fundamental problems for sensor networks require a co-design approach, i.e. components of communication protocol of different levels should be considered simultaneously.

One of important and crucial elements for the successful design of sensor networks is the algorithmic component which defines the way in which the sensors and other elements cooperate in order to build a well-functioning, dependable and robust system. The number of problems is manifold. Below we touch only a few of them.

Energy Aware Design. One of the crucial problems for sensors is energy supply: improvement over capacity of batteries turn out to be possible, but relatively slow. Consequently, a designer of a sensor network has to be aware about technical feasibility regarding energy usage. The most problematic issue is relatively very high energy usage for radio communication; it concerns both broadcasting data and passive monitoring of the radio channel.

Exchanging exhausted batteries is limited to some application scenarios; in quite many cases we apply sensors at places that are hardly accessible, or even hidden (for instance, it might be the case for sensors monitoring pollution by a chemical factory, otherwise the sensing devices could be damaged or manipulated).

M. Kutyłowski et al. (Eds.): ALGOSENSORS 2007, LNCS 4837, pp. 1–5, 2008.
© Springer-Verlag Berlin Heidelberg 2008

Due to the reasons described, it is very important to design algorithmic schemes that reduce energy usage. This may concern amount of computations performed by a sensor, duration of sensing activities, however the key issue is to reduce amount of communication and to optimize transmission parameters. For instance, even if a destination of data sent by a sensor is in the transmission range of the sensor, it might be useful to send data via a few intermediate sensors - each time sending a short range message to the next sensor on a path to the destination point. This should save energy, since the amount of energy required to send on distance d is at least of order d^2. However, such an approach generates a few new problems: one of them might be congestion of messages passing through a node. Even if each of these messages requires a low amount of energy, the number of messages and necessity to monitor the radio channel for a longer time may lead to exhausting energy at this point.

Energy saving is a topic that goes across all layers of design, however a few challenges can be addressed already during hardware design. For instance, protecting data contents could be tuned to special needs of sensor networks: relatively weak mechanisms could be implemented directly in hardware so that data are encoded and decoded fast and almost no communication overhead arises.

Network Architecture. At an early stage of research on sensor networks it has been assumed that the sensors compose a fully autonomous communication network that works in an ad hoc mode. This was motivated by applications for which a sensor network should be as independent as possible (networks for emergency and military purposes).

On the other hand, in quite many practical situations data can be gathered by strong units having external power supply and therefore not so limited in communication. Furthermore, they can use broad communication links (wired and wireless) to process data to their final destinations. They can also take some responsibility for configuring the sensors and run-time administration.

Even if such an architecture simplifies design of sensor networks, it implies new challenges. An example of such a challenge is presence of more than one strong unit in the communication range of a sensor. In this case the strong units can partition the sensors between themselves, but this is a solution that might be inefficient for many reasons. In order to build a robust and reliable system with overlapping coverage areas it is necessary to share sensors between adjacent strong units. How to coordinate their work so that no conflicts and waste of resources occur?

Channel Access, Interference. One of the key technical problems for sensor networks is to organize access to its shared radio channel. A simultaneous transmission by more than one sensor may result in a collision so that no message will be received, and therefore transmission time will be wasted. This is a problem since energy and transmission time are limited resources.

Interference between signals can occur in a somewhat unpredictable way, it depends on many issues that are hard to be modeled mathematically. For example, the unit graph model does not describe exactly propagation of radio signals:

A propagation range of a sensor's transmitter cannot be stated as a single value, a more realistic model concerns probability of recognizing a message as a function of distance between the transmitter and the receiver. Reflection of signals, lack of equidirectional transmission, peculiarities of interferences between different frequencies, radio traffic in other frequency bands, noise in the radio channel coming from the environment, and many further issues make it hard to predict exactly what kind of problems will occur after deploying a sensor network.

Even if communication architecture of a sensor network is simple and can be modeled as (overlapping) single hop networks with strong units gathering data and coordinating all activities in the network, many uneasy problems arise. A good example is a network that has to warn about certain conditions (like increasing temperature). The main issue is that we do not know who, where and when will need to report some event. Moreover, if such an event occurs there might be many sensors that would like to transmit at the same time. Round Robin strategy might be a bad choice in this case, especially if reaction time of the network is the key issue.

New Features of Communication Complexity. Certainly, complexity measures such as communication complexity are very helpful for designing algorithms for sensor networks. However, the communication volume does not capture well all specific issues of sensor networks. A more adequate complexity measure may be based on active communication time (expressed for instance in the number of bits that can be sent or received in this time). However, we have to consider also the number of activity periods of a receiver/transmitter, since activating it requires both time and energy - and for this reason sending just a few bits can be very costly regardless of the small volume sent. The second issue is exploiting asymmetry: energy consumption for sending and and for receiving messages are comparable, still the difference can be used to optimize the energy usage. In particular, we could look for communication schemes reduce transmission time of the sensors and assign transmission chores to strong units.

Last not least, in many situations we should treat a sensor network as a whole and consider communication like an activity that uses distributed resources and in this way decays the capabilities of the network as a whole. In some settings it is better to sacrifice a single sensor (due to energy exhaustion) and save energy of other sensors than to use energy of all sensors in an egalitarian way. New complexity measures have to capture somehow the global resources usage of a sensor network and not its individual units.

Preprocessing of Data. Data aggregation is a process that must take into account limitations of communication and local computing capabilities of a sensor. The simplest strategy of forwarding all data sensed resolves the problem of data analysis by the sensor, but leads to communication congestion and waste of energy. Therefore, data should be preprocessed by a sensor and only essential information should be transmitted by the sensor.

The problem is that the sensor has typically only local knowledge and a limited knowledge from its neighborhood. For the sake of energy preserving a sensor's receiver should not be switched on all the time, so inevitably some messages broadcasted could be unnoticed by the sensor, despite that the sensor was in the proper range.

This problem is particularly visible when we concern warning function of sensors. Detecting anomalies (which are the most crucial ways of detecting critical situations) is not always equivalent with detecting certain numeric values, sometimes it is their gradient or an abnormal pattern of values in some area. How to detect abnormal patterns by a set of sensors without large communication is an interesting question that involves issues of game theory, distributed algorithms and communication complexity.

Heterogeneous Networks. In many practical applications we could share the basic infrastructure composed by sensors. This helps to reduce deployment costs and compose many virtual sensor networks out of the same set of devices. Inevitably, we may come to a situation when a virtual sensor network consists of devices that are deployed by more than one provider. Consequently, sensor type and their configuration may differ, sometimes quite significantly.

Since sensors are weak devices, it might be an uneasy for them to emulate different configurations in order to serve the needs of diverse virtual networks. Therefore, algorithms and protocols designed for sensor networks should to certain degree disregard particular features of the sensors - a good algorithm should be as independent of the features of the sensors as possible. At the same time, it should make use of any additional features available at certain nodes of the network.

Standard algorithm design is not focused on such a situation: usually it is assumed that in a distributed system all network units have comparable properties. Therefore most of the algorithms for distributed systems have to be redesigned or at least reconsidered before being deployed for sensor networks.

Network Evolution. Heterogenuity may arise for yet another important reason: after some time new sensors become deployed in order to upgrade its functionality or to replace malfunctioning or dead units. Location, functionality and purpose of strong units may change as well. So finally we are faced with many compatibility problems: new devices should work well within an old network, however, they have to take advantage of newer technology in order to improve the network performance.

The problem is a more difficult than in the classical networks. Sometimes it is impossible to exchange fully the old units. Sometimes we share sensors with another provider who can decide to make a replacement without our accord. The problem is that communication protocols for sensor networks are not that stable as for the classical networks, so a replacement may have profound consequences.

Since we have to assume that changes within the network may be frequent and quite significant, and involve dramatically different hardware, we have to design self-organization paradigms yielding algorithms that are robust to these changes. We have to design algorithmic schemes (and not only data framework!) that support upwards and downwards compatibility.

Dependability. There are many factors that may cause faults of the sensors. Extreme physical conditions may cause both transient and permanent failures. Internal failures may occur as well. Furthermore, the sensors might be manipulated or even damaged on purpose. Finally, some errors might occur due to design faults. Sometimes design bugs cannot be corrected - the faulty code may be written in a ROM memory. Sensor network algorithms should take into account all these problems.

So far there are not many algorithms focused on fault conditions. Even worse, majority of algorithms break down at the moment when faults occur. Sometimes they loose efficiency, hang on, or even yield wrong results.

There are many algorithmic challenges in this area. One of the hard problems is to redesign self-organization algorithms so that they yield proper results even if different sensors may have inconsistent view of the state of the shared radio channel. For example, it may happen that some sensors receive a message and can interpret it correctly, while other sensors find it unreadable. Classical algorithms assume that a message is either scrambled or accessible for all sensors that haaave their receivers switched on.

Trustworthiness. For some application areas it is necessary to guarantee that the data delivered by sensor networks are trustworthy. For instance, if a sensor network has to monitor environment pollution, then the data provided by the network should serve as a indisputable evidence in a court. So it is necessary to design some mechanisms that insure the data origin, time, location, and lack of manipulation. On the other hand, it is quite easy to find a sensor and replace it, or manipulate unprotected radio messages.

Contemporary cryptography provides a full range of cryptographic tools for authentication and protecting data against manipulations and unauthorized reading. The problem is that the classical methods are quite "heavy" and not well suited for sensor networks. The problems are communication overhead, computational resources necessary for performing cryptographic operations, key management and quite high probability of compromising some nodes.

New algorithms and schemes should take into account these issues: they should reduce communication and computation demands and become safe even if a certain fraction of nodes become compromised. This seems to be hardly possible to meet all these demands, but the challenge is not to secure each single node (which might be hard), but to guarantee smooth operation of the network as a whole. Lack of security of single units should be compensated by "joint behavior" of the network.

Topology and Routing in Sensor Networks*

Sándor P. Fekete and Alexander Kröller

Department of Computer Science
Braunschweig University of Technology
D-38106 Braunschweig, Germany
{s.fekete,a.kroeller}@tu-bs.de

Abstract. At ALGOSENSORS 2004 we presented a first algorithm to detect the boundary in a dense sensor network. This has started a new field of research: how to establish topology awareness in sensor networks without using localization. Three years later, at ALGOSENSORS 2007, we present a number of further results.

After discussing issues of distance estimation and computation of co-ordinates, we give an overview over the boundary recognition problem and show a new approach to solving it. Then we show how to use the boundaries for higher-order topology knowledge; the outcome is a graph that describes the network topology on a very high level, while being small enough to be distributed to all nodes. This allows every node in the network to obtain knowledge about the global topology. Finally, we show how to use these structures for efficient routing.

1 Introduction

In recent time, the study of wireless sensor networks (WSN) has become a rapidly developing research area. Typical scenarios involve a large swarm of small and inexpensive processor nodes, each with limited computing and communication resources, that are distributed in some geometric region; communication is performed by wireless radio with limited range. Upon start-up, the swarm forms a decentralized and self-organizing network that surveys the region.

From an algorithmic point of view, these characteristics imply absence of a central control unit, limited capabilities of nodes, and limited communication between nodes. This requires developing new algorithmic ideas that combine methods of distributed computing and network protocols with traditional centralized network algorithms. In other words: how can we use a limited amount of strictly local information in order to achieve distributed knowledge of global network properties? As it turns out, making use of the underlying geometry is essential.

In this paper, we give an overview of topics and results discussed and presented during an invited presentation at ALGOSENSORS 2007. Section 2 deals with location awareness; we start by describing distance estimation without the use of special hardware; this is followed by a discussion of the limitations of the

---

* This invited survey article is based in parts on excerpts from our articles [6,7,8,15].

M. Kutyłowski et al. (Eds.): ALGOSENSORS 2007, LNCS 4837, pp. 6–15, 2008.

computation of node coordinates. Section 3 gives a description of our approach to topology recognition, consisting of boundary recognition and topological clustering, which has been turned into a video, based on large-scale simulation. Section 4 describes some new approaches to routing.

2   Location Awareness

One of the key problems in sensor networks is to let nodes know their location, for example, by storing coordinates w.r.t. a global coordinate system. Unless all nodes are equipped with special localization devices (e.g., GPS/Galileo), there needs to be an algorithm that computes positions based on information available to the network.

2.1   Distance Estimation

Practical localization algorithms often use connectivity information enriched with distance estimates for adjacent nodes [17]. Note that the corresponding decision problem is NP-hard [1].

Various ways to measure distance exist. Examples include the transmission time-of-flight over a wireless channel, the latency of infrared communication, or the strength of a wireless signal that decreases with distance. Good approaches have an average error of about 10–20% of the maximal communication range.

Our approach does not rely on special hardware or node capabilities. Assuming the probability of successful communication decreases with increasing distance, the expected fraction of a node's neighbors that it shares with an adjacent node defines a monotonically decreasing function that can be inverted, resulting in a distance estimator based on this fraction. All that is required is the ability to exchange neighbor lists and a model of communication characteristics.

We assume that nodes are uniformly distributed over the plane, with density δ. That is, the expected number of nodes in a region $A \subset \mathbb{R}^2$ of area $\lambda(A)$ equals $\delta\lambda(A)$. The neighborhood N_i of a node i depends on communication characteristics, which are modelled by an appropriate communication model. We focus on symmetric models only, i.e., $i \in N_j$ iff $j \in N_i$. The model is a probability function $p(d)$ that defines the probability that two nodes i and j with distance $d = \|i - j\|$ are connected. Hence, the expected size of a neighborhood is $\mathbb{E}[|N_i|] = \delta \int_{\mathbb{R}^2} p(\|x\|)dx$ for all nodes i.

We want to estimate the distance of i and j by counting how many of i's neighbors are shared with j. The expected size of this fraction is

$$f_p(d) := \mathbb{E}[|N_i \cap N_j|/|N_i \setminus \{j\}|] \qquad (1)$$
$$= \frac{\int_{\mathbb{R}^2} p(\|x\|)p(\|x - (d,0)^\mathsf{T}\|)dx}{\int_{\mathbb{R}^2} p(\|x\|)dx},$$

where $d = \|i - j\|$. If f_p^{-1} exists, two nodes i and j can exchange their neighbor lists, compute the shared fraction $\varphi_{i,j} = |N_i \cap N_j|/|N_i \setminus \{j\}|$ and estimate their

Table 1. Average estimation errors for different densities

Scaled density $\pi\delta$	5	8	10	15	20	40	80
Error (inner nodes)	.225	.183	.165	.137	.120	.087	.062
Error (boundary nodes)	.257	.201	.182	.154	.135	.101	.077

distance as $f_p^{-1}(\varphi_{i,j})$. Note that $\varphi_{i,j}$ and $\varphi_{j,i}$ may be different, so some additional tie breaking or averaging scheme must be used.

There is an elegant way to implement this approach for practical purposes, as proposed by Buschmann et al. [3]: Instead of f_p^{-1}, a small discrete value table of f_p is stored in the nodes, and the estimate is done by reverse table lookup. This even works for p or f_p obtained by numerical or field experiments, and it can be implemented using only integer arithmetic.

A widely used model for radio networks is the Unit Disk Graph (UDG), where two nodes i and j are connected by a link iff $\|i - j\| \leqslant 1$.

For UDGs, the estimated neighborhood fraction (1) is $f : [0,1] \to [0,1]$ with

$$f(d) = \frac{2}{\pi}\left(\arctan(\tfrac{d}{2}) - \tfrac{d}{2}\sqrt{1 - (\tfrac{d}{2})^2}\right) . \tag{2}$$

f^{-1} exists, but unfortunately we lack a closed formula for it. Instead, we approximate f^{-1} by its Taylor series about $f(0) = 1$. Here, we use the 7th-order Taylor polynomial

$$t_7(\varphi) = -\frac{\pi}{1! \, 2}(\varphi - 1) - \frac{\pi^3}{3! \, 2^5}(\varphi - 1)^3$$
$$- \frac{13\pi^5}{5! \, 2^9}(\varphi - 1)^5 - \frac{491\pi^7}{7! \, 2^{13}}(\varphi - 1)^7. \tag{3}$$

We do not use a higher order because evaluating the polynomial on practical embedded systems would become numerically unstable.

To evaluate the UDG estimator's performance, we ran some simulations. Table 1 shows their results. The first row contains the expected size of a neighborhood, without boundary effects. For UDGs, this is $\pi\delta$. Furthermore, the average errors are reported. The error is relative to the communication range, which is the common measure for distance estimators. The average is taken separately for two classes of links: for "inner" links, the communication ranges of both end-nodes are fully contained in the network region. For "boundary" links, both end-nodes lie at most 1 from a straight boundary. This separation has two benefits: First, the estimate in its current form focuses on the inner links only, and second, it removes the dependency on the network region's shape from the evaluation.

One can see how our approach already reaches the desired accuracy of 20% for an average neighborhood size of just eight, and gets even better for larger neighborhood size.

Fig. 1. Left: Example network with marked anchor nodes. Right: Result of Ad-Hoc Positioning [18].

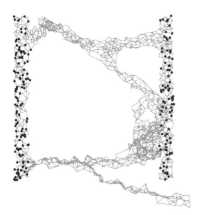
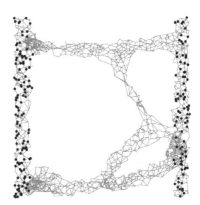

Fig. 2. Left: Result of Robust Positioning [20]. Right: Result of N-Hop Multilateration [21].

2.2   Coordinates

When trying to understand the network structure, a seemingly natural approach is to determine the underlying node coordinates, based on trigonometric computations that use node distances. A variety of methods have been proposed, including [18, 20, 21, 19]. Unfortunately, these methods show a number of problems in the presence of errors and large numbers of nodes: see our Figures 1, 2, 3 for an example with 2200 nodes, with a subset of "anchor" nodes that know their precise coordinates marked in black; node distances have an a random error with a standard deviation of 1%.

Quite clearly, the resulting embeddings do not only produce imprecise node coordinates; more importantly, they do not accurately reflect the network topology. This means that in the context of exploiting the network topology for

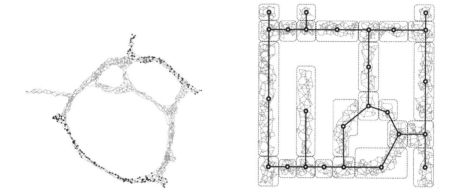

Fig. 3. Left: Result of Anchor-Free Localization [19]. Right: Our alternative: clusters and cluster graph.

purposes such as routing, computing more or less accurate coordinates is indeed a red herring; as will show in the following, a suitable alternative is to consider topological clustering, as motivated in Figure 3. See our article [15] for more technical details.

3   Topology Awareness

3.1   Boundary Recognition

Recognizing the network boundary is vital for detecting objects entering or leaving the monitored area or events that affect the network structure. Boundary detection is also a stepping stone towards organizing the network.

In the setting described above, we first proposed the problem in [9], using a probabilistic setting; see [5] for a refined approach. A different approach for our problem was suggested by Funke [11], and requires a particular boundary structure and sufficient density; see [12] for details. Another approach was proposed by Wang et al. [22]. Our method described in [14] yields deterministically provable results for any kind of boundary structure. We assume that the communication graph is a $\sqrt{2}/2$-quasi unit disk graph, a generalization of unit disk graphs first introduced by [2].

See Figure 4 for a geometric representation of a graph called a *flower*. Non-neighboring nodes have a minimum distance, so an independent node set requires a certain amount of space in the embedding. On the other hand, the space surrounded by a cycle is limited. This packing argument allows conclusions about the relative embedding of nodes. Applying a similar argument repeatedly, the central nodes can deduce that they lie inside of the outer cycle.

Because flowers are strictly local structures, they can be easily identified by local algorithms. In the shown example network of 60,000 sparsely connected nodes, our procedure identified 138 disjoint flowers; a single suffices for the second stage of our algorithm.

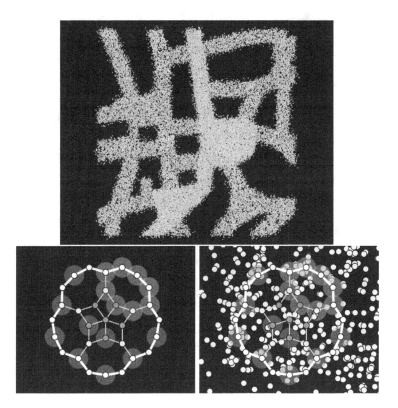

Fig. 4. Top: A sensor network, obtained by scattering 60,000 nodes in a of street network. Bottom left: A flower subgraph, used for reasoning about boundary and interior of the network. Right: A flower in the context of the network shown above.

In this second stage (cf. Figure 5), we augment flowers by adding extensions to their outer cycles, such that insideness can still be proven for all contained nodes. By repeating a local search procedure, the flowers grow to enclose more and more nodes and merge together, eventually leading to a single structure that contains most of the network.

3.2 Topological Clustering

We use the identified boundaries to construct a topological clustering; see Figure 6. By considering the hop count from the boundary, we get a shortest-path forest. Nodes that have almost the same distance to several pieces of the boundary form the medial band of the region. Nodes close to three different boundary portions form vertices of the medial band, or medial vertices. After identifying a set of medial vertices, we also know their distance from the boundary. This makes it easy to grow the corresponding intersection cluster to just the right

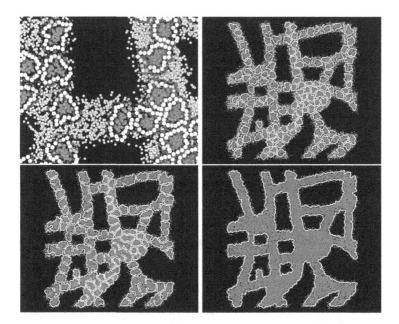

Fig. 5. Top left: A set of expanding flowers in a network. Top right: A later stage of the flower expansion. Bottom left: Flowers merging together. Bottom right: The result of boundary recognition.

size. Thus, we can identify all intersection clusters in the network. Finally, parts of the network adjacent to intersection clusters give rise to street clusters.

In the end, we have structured the network into a natural set of clusters that reflect its topology. This makes it is possible to perform complex tasks, such as tracking and guiding, based on purely local operations.

3.3 Our Video

Using our toolbox SHAWN [16] for the simulation of large and complex networks, we have produced a video that illustrates two procedures for dealing with the above algorithmic challenges: one identifies the boundaries of the network; the other constructs a clustering that describes the network topology. For more technical details of the underlying algorithmic side, see our paper [14]. Our software is freely available at www.sourceforge.net/projects/shawn.

The video starts by describing sensor nodes and their deployment. Following an introduction of the algorithmic challenge, the next scene illustrates the problem of boundary recognition. Next is the concept of flowers and the way they allow local, deterministic reasoning about nodes lying in the interior of the network. Their extension by augmenting cycles is demonstrated in the following scene, leading to full-scale boundary recognition. The next part introduces the medial band and the recognition of medial vertices. In the final sequence, this leads to the construction of clusters.

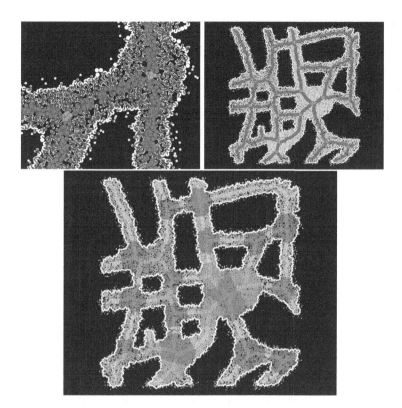

Fig. 6. Top left: The shortest-path forest defines the medial band. Top right: Medial band and medial vertices. Bottom: The resulting intersection and street clusters.

4 Routing

Once the above cluster structure has been extracted, it is straightforward to use it for network tasks such as routing and tracking: on the large scale, use the cluster graph G, either on a global scale (if G is small enough to be available all over the network), or for local handover between clusters (if G is not available); within each cluster, use virtual coordinates that arise (a) by hop distance from the cluster boundary (b) by hop distance along the cluster boundary. Using more refined geometric structures and properties, it is possible to achieve good routing results. See the forthcoming Ph.D. thesis [13] of the second author for details.

A related problem is to find approximately shortest paths in a clustered sensor network, without having access to node coordinates or a detailed routing table. For the case of a subdivision into convex clusters, our group has developed a memory-efficient, local algorithm that guarantees a 5-approximation of the shortest routing path. See the masters thesis by Förster [10], and our related paper [4] for details.

Acknowledgments

Parts of this work presented in this survey were based on a long-standing, very fruitful collaboration with Stefan Fischer, Dennis Pfisterer and Carsten Buschmann from University of Lübeck. We thank our colleagues for making it possible. The work by Alexander Kröller was part of project "SwarmNet" within the DFG priority programme SPP 1126, "Algorithms for large and complex networks", grant numbers FE 407/9-1 and FE 407/9-2.

References

1. Aspnes, J., Goldenberg, D., Yang, Y.R.: On the computational complexity of sensor network localization. In: Nikoletseas, S.E., Rolim, J.D.P. (eds.) ALGOSENSORS 2004. LNCS, vol. 3121, pp. 32–44. Springer, Heidelberg (2004)
2. Barrière, L., Fraigniaud, P., Narayanan, L.: Robust position-based routing in wireless ad hoc networks with unstable transmission ranges. In: DIALM 2001, Rome, Italy, pp. 19–27. ACM Press, New York, NY, USA (2001)
3. Buschmann, C., Hellbrück, H., Fischer, S., Kröller, A., Fekete, S.P.: Radio propagation-aware distance estimation based on neighborhood comparison. In: Langendoen, K., Voigt, T. (eds.) EWSN 2007. LNCS, vol. 4373, pp. 325–340. Springer, Heidelberg (2007)
4. Fekete, S.P., Förster, K.-T., Kröller, A.: Using locally optimal paths for memory-efficient routing in sensor networks, manuscript (2007)
5. Fekete, S.P., Kaufmann, M., Kröller, A., Lehmann, N.: A new approach for boundary recognition in geometric sensor networks. In: Proceedings 17th Canadian Conference on Computational Geometry, pp. 82–85 (2005)
6. Fekete, S.P., Kröller, A.: Geometry-based reasoning for a large sensor network. In: Proc. 22nd Annual ACM Symposium on Computational Geometry, pp. 475–476 (2006), http://compgeom.poly.edu/acmvideos/socg06video/index.html
7. Fekete, S.P., Kröller, A., Buschmann, C., Fischer, S.: Geometric distance estimation for sensor networks and unit disk graphs. In: Proceedings 16th International Fall Workshop on Computational Geometry (2006)
8. Fekete, S.P., Kröller, A., Pfisterer, D., Fischer, S.: Algorithmic aspects of large sensor networks. In: Proceedings Mobility and Scalability in Wireless Sensor Networks, pp. 141–152. CTI Press (2006)
9. Fekete, S.P., Kröller, A., Pfisterer, D., Fischer, S., Buschmann, C.: Neighborhood-based topology recognition in sensor networks. In: Nikoletseas, S.E., Rolim, J.D.P. (eds.) ALGOSENSORS 2004. LNCS, vol. 3121, pp. 123–136. Springer, Heidelberg (2004)
10. Förster, K.-T.: Clusterbasierte Objektüberwachung in drahtlosen Sensornetzwerken: Lokal optimale Wege in der Ebene und ihre Anwendung in Sensornetzwerken. Master's thesis, Braunschweig University of Technology, Braunschweig (2007)
11. Funke, S.: Topological hole detection in wireless sensor networks and its applications. In: Proc. DIALM-POMC, pp. 44–53 (2005)
12. Funke, S., Klein, C.: Hole detection or: how much geometry hides in connectivity? In: Proc. 22nd ACM Symp. Comput. Geom., pp. 377–385 (2006)
13. Kröller, A.: Algorithms for topology-aware sensor networks. PhD thesis, Braunschweig University of Technology, Braunschweig (2007)

14. Kröller, A., Fekete, S.P., Pfisterer, D., Fischer, S.: Deterministic boundary recognition and topology extraction for large sensor networks. In: Proc. 17th ACM-SIAM Sympos. Discrete Algorithms, pp. 1000–1009 (2006)
15. Kröller, A., Fekete, S.P., Pfisterer, D., Fischer, S., Buschmann, C.: Koordinatenfreies Lokationsbewusstsein. Information Technology 47, 70–78 (2005)
16. Kröller, A., Pfisterer, D., Buschmann, C., Fekete, S.P., Fischer, S.: Shawn: A new approach to simulating wireless sensor networks. In: DASD 2005. Proceedings Design, Analysis, and Simulation of Distributed Systems, pp. 117–124 (2005)
17. Langendoen, K., Reijers, N.: Distributed localization in wireless sensor networks: A quantitative comparison. Computer Networks 43(4), 499–518 (2003)
18. Niculescu, D., Nath, B.: Ad-hoc positioning system (APS). In: Proceedings of the 44th IEEE Global Telecommunications Conference (GLOBECOM 2001), pp. 2926–2931 (November 2001)
19. Priyantha, N.B., Balakrishnan, H., Demaine, E., Teller, S.: Anchor-free distributed localization in sensor networks. In: Proc. 1st Intl. Conf. on Embedded Networked Sensor Systems (SenSys), Los Angeles, CA, pp. 340–341. ACM Press, New York (2003)
20. Savarese, C., Rabay, J., Langendoen, K.: Robust positioning algorithms for distributed ad-hoc wireless sensor networks. In: Proc. USENIX Technical Ann. Conf., Monterey, CA (2002)
21. Savvides, A., Park, H., Srivastava, M.B.: The bits and flops of the n-hop multilateration primitive for node localization problems. In: Proc. 1st ACM Intl. Workshop on Sensor Networks and Applications (WSNA), Atlanta, GA (2002)
22. Wang, Y., Gao, J., Mi

Codes for Sensors: An Algorithmic Perspective[*]

João Barros

Instituto de Telecomunicações
Departamento de Ciência de Computadores
Faculdade de Ciências da Universidade do Porto
Rua do Campo Alegre, 1021/1055, 4169-007 Porto, Portugal
barros@dcc.fc.up.pt

Abstract. Sensing, processing and transmitting data are arguably the key activities of common nodes in a wireless sensor network once embedded in a physical process unfolding in time and space. To process the data and transmit it reliably and efficiently over noisy links along the network, sensor nodes require codes that are capable of exploiting the natural correlation of the gathered data and of combating the impairments caused by noisy communication channels.

Once we define reasonable models for the information sources and the communication channels, information theory offers powerful tools to study the ultimate performance limits for any coding scheme designed for this class of communication and computation systems. To illustrate this observation, we start by modeling the sensor network as a set of multiple correlated sources that are observed by partially cooperating encoders and transmitted over a network of independent channels. Based on this formulation, we are able to characterize the network capacity, i.e., the exact conditions on the sources and the channels under which there exist codes for reliable communication with the data collection point. An important conclusion to be drawn from our proofs, is that for a large (and arguably most relevant) class of sensor networks, separate data compression and error correction codes provide an optimal system architecture.

The proofs also offer hints on how to construct practical algorithms for distributed compression and joint inference of correlated data collected by hundreds of sensor nodes. After showing that the optimal decoder based on minimum mean square estimation (MMSE) is unfeasible – its complexity grows exponentially with the number of nodes – we present a two-step "scalable" alternative: (1) approximate the correlation structure of the data with a suitable factor-graph, and (2) perform joint source/channel decoding on this graph using the sum-product algorithm. Based on this general approach, which can be applied to sensor networks with arbitrary topologies, we give an exact characterization of the decoding complexity, as well as optimization algorithms for finding optimal factor trees under the Kulback-Leibler criterion. Finally, we are able to show how these ideas can also be used for distortion-optimized index assignments for low-complexity distributed quantization, and source-optimized hierarchical clustering.

---

[*] Keynote speech.

M. Kutyłowski et al. (Eds.): ALGOSENSORS 2007, LNCS 4837, pp. 16–17, 2008.

Acknowledgements

The author gratefully acknowledges many useful discussions with Sergio D. Servetto, Gerhard Maierbacher and Michael Tuechler.

Efficient Sensor Network Design for Continuous Monitoring of Moving Objects*

Sotiris Nikoletseas and Paul G. Spirakis

CTI and Univ. of Patras, Greece,
{nikole,spirakis}@cti.gr

Abstract. We study the problem of localizing and tracking multiple moving targets in wireless sensor networks, from a network design perspective i.e. towards estimating the least possible number of sensors to be deployed, their positions and operation chatacteristics needed to perform the tracking task. To avoid an expensive massive deployment, we try to take advantage of possible coverage ovelaps over space and time, by introducing a novel combinatorial model that captures such overlaps.

Under this model, we abstract the tracking network design problem by a combinatorial problem of covering a universe of elements by at least three sets (to ensure that each point in the network area is covered at any time by at least three sensors, and thus being localized). We then design and analyze an efficient approximate method for sensor placement and operation, that with high probability and in polynomial expected time achieves a $\Theta(\log n)$ approximation ratio to the optimal solution. Our network design solution can be combined with alternative collaborative processing methods, to suitably fit different tracking scenaria.

1 Introduction

1.1 Problem Description

We wish to solve the problem of *localizing and continuously tracking mobile objects* moving in a domain described by a set of set of three-dimensional curves, S, over a period of time T i.e. we want the wireless sensor network to be able to detect the position of any moving object, at any time t in T. We allow multiple targets that arise in the network area at random locations and at random times. The movement of each target can follow an arbitrary but continuous path i.e. we disallow the target to instantaneously "jump" to another location; still, we can handle such discontinuities as multiple targets.

In our setting, the set S of 3D curves is the set of possible trajectories of objects moving for some time within the period T. Such a moving object might follow a part of a curve in S, and possibly arrive to an intersection of curves and then follow another curve.

---

* This work was partially supported by the IST/FET/Global Computing Programme of the European Union, under contact number IST-2005-15964 (AEOLUS).

M. Kutyłowski et al. (Eds.): ALGOSENSORS 2007, LNCS 4837, pp. 18–31, 2008.

We can assume that each such curve in S is specified (for $t = 0, \ldots, T$) by an analytic equation (e.g. in x, y, z coordinates). Thus, we can compute a curve (route) *piece* τ_{ij} for the curve (route) R_i, via some criterion (e.g. to split R_i into pieces of equal length). We, then, wish to deploy some sensors (either standing or even moving for some time) in some (initial) places within S. A basic demand for such a deployment consists of the following rule:

(Rule R) For every point $\overrightarrow{p} = (x, y, z)$ in S and for every $t \in [0, T]$ there are at least three sensors, active at time t, whose sensing range includes the point $\overrightarrow{p}$. Here, for a sensor σ operating at time t, its *sensing range* is a sphere of some radius $R_\sigma(t)$ and with center the position of σ at time t.

Note that rule R guarantees localization of any moving target in S at any time t in T, via triangulation.

We follow a *network design approach* to this problem. Let us assume, hypothetically, that we could have an abundance of sensors, each characterized by an initial position, an operation period, and an initial available energy (battery) that allows a particular implementation of sensing ranges during the operation of the sensor. Then, the decision to actually select one sensor (with the initial position, operating period and available energy) has a certain *cost*. Our goal is to be able to select a subset of sensors that implement rule (R) and is of nearly minimal cost.

The problem we study is related (but different) to the problems of network coverage and tracking. In fact, we extend the well-studied coverage problems by being able to track the moving path, and by also taking time into account. On the other hand, to reduce the energy dissipation and overhead of our tracking solution, we avoid some of the collaborative information processing components (like which nodes should sense, which have useful information and should communicate, which should receive information and how often). Thus, we are not dealing directly with queries of the type "how many targets are in a certain region during a certain time interval". Still, our solution performs the collaborative processing tasks (triangulation by at least three sensors) that allow localization of the targets as they move in the network. Also, we discuss how alternative collaborative processing methods can be combined with our approach to provide full tracking.

1.2 Our Contribution

Our approach indeed tries to avoid the expensive, massive placement of all the time functioning sensors all over the monitoring area, *by exploiting possible overlaps of routes* at certain places in the network area, using sensors that can simultaneously monitor pieces of several (nearby) routes during their operation, or even exploit sensors on top of certain moving objects of our own that wander in the maze of routes. Thus, we somehow "multiplex" (both over space and time) the use of deployed sensors, since our approach identifies overlaps over space and time and thus deploys fewer sensors compared to the trivial approach.

In fact, the analysis of our method shows that *it is very efficient*, especially in very large domains, both with respect to computational time and the deployment cost, since it finds in polynomial expected time a deployment solution which approximates the optimal solution within a logarithmic factor. Thus, we avoid expensive dense deployment of sensors, where the information about the target is simultaneously generated by multiple sensors; we instead achieve a low cost solution (by removing unnecessary redundancy) that still keeps tracking accuracy at high levels.

To be able to handle overlaps, *we propose a novel combinatorial model for possible routes* in the network domain. This model, although abstract, captures several of the technical specifications of real sensor devices, such as the energy spent as a function of the transmission range, the ability to vary this range to save energy or increase connectivity, the ability of sensors to employ power saving (sleep-awake) schemes to save energy etc. This combinatorial model allows us to reduce the tracking problem to a variation of a combinatorial problem of set covering (in particular to "at least 3 cover", i.e. having each point of the domain covered at any time by at least 3 sensors, and thus being localized). We feel that this combinatorial model is of independent interest and can (itself or its variatios) be used in modeling other problems as well.

1.3   Related Work and Comparison

As discussed in the problem definition part, the problem we study is relevant to network coverage and tracking, that we discuss below.

Coverage. Sensor deployment strategies play a very important role in providing better QoS, which relates to the issue of how well each point in the sensing field is covered. However, due to severe resource constraints and hostile environmental conditions, it is nontrivial to design an efficient deployment strategy that would minimize cost, reduce computation, minimize node-to-node communication, and provide a high degree of area coverage.

Several deployment strategies have been studied for achieving an optimal sensor network architecture. Dhilon et al. ([4]) propose a grid coverage algorithm that ensures that every gridpoint is covered with a minimum confidence level. They consider a minimalistic view of a sensor network by deploying a minimum number of sensors on a grid that would transmit a minimum amount of data. Their algorithm is iterative and uses a greedy heuristic to determine the best placement of one sensor at a time. It terminates when either a preset upper limit on the number of sensors is reached or sufficient coverage of the gridpoints is achieved. However, the algorithm assumes line of sight of the target and the sensor. Also, since a complete knowledge of the terrain is assumed, the algorithm is not very applicable in cluttered environments, such as interior of buildings, because modeling obstacles becomes extremely difficult in those scenarios. Finally, a main difference of this approach (and in fact the other coverage methods as well) with ours, is that we explicitly take time into account.

In contrast to static sensor networks, nodes in mobile sensor networks are capable of moving in the sensing field. Such networks are capable of self-deployment starting from an initial configuration. The nodes would spread out such that coverage in the sensing field is maximized while maintaining network connectivity. A potential field-based deployment approach using mobile autonomous robots has been proposed to maximize the area coverage ([14]). Clearly, the assumptions of this method (and the one described below) are different to ours, since we do not use sensor mobility as an algorithmic design element (although we handle mobility when it appears).

Similar to the potential field approach, a sensor deployment algorithm in the presence of mobility based on virtual forces has been proposed in [19] to increase the coverage after an initial random deployment. A sensor is subjected to forces, which are either attractive or repulsive in nature. In this approach, obstacles exert repulsive forces, while areas of preferential coverage (sensitive areas where a high degree of coverage is required) exert attractive forces, and other sensors exert attractive or repulsive forces. A hard threshold distance is defined between two sensors to control how close they can approach each other.

Other interesting network coverage approaches are discussed in the book chapter by A. Ghosh and S. Das in [16].

Tracking. Our method follows and extends the well-established line of research for a network architecture design for centralized placement/distributed tracking (see e.g. the book [16] for a nice overview). According to that approach, optimal (or as efficient as possible) sensor deployment strategies are proposed to ensure maximum sensing coverage with minimal number of sensors, as well as power conservation in sensor networks.

Centralized Approaches. In one of the methods ([3]), that focuses on deployment optimization, a grid manner discretization of the space is performed. Their method tries to find the gridpoint closest to the target, instead of finding the exact coordinates of the target. In such a setting, an optimized placement of sensors will guarantee that every gridpoint in the area is covered by a unique subset of sensors. Thus, the sensor placement problem can be modeled as a special case of the alarm placement problem described by Rao [15]. That problem is the following: given a graph G, which models a system or a network, one must determine how to place alarms on the nodes of G so that any single node fault can be diagnosed. It has been shown in [15] that the minimal placement of alarms for arbitrary graphs is an NP-complete problem. Clearly, their problem is easier than ours, since they relax the requirement to find exact coordinates of moving objects by just finding the nearest gridpoint. Since their problem is computationally difficult, this implies the inherently high complexity of our problem. Another indication of the hardness of the problem is the fact that, as shown in [2], the localization problem is NP-hard in sparse wireless sensor networks.

Also, our problem is related but different to the following other well known approach that focuses on power conservation: in [7] sleep−awake patterns for each sensor node are obtained during the tracking stage, to obtain power efficiency. The network operates in two stages: the surveillance stage during the absence of any event of interest, and the tracking stage, in response to the presense of moving targets. Each sensor initially works in the low-power mode when there are no targets in its proximity. However, it should exit the low-power mode and be active continuously for a certain amount of time when a target is sensed, or even better, when a target is shortly about to enter. Finally, when the target passes by and moves farther away, the node should decide to switch back to the low-power mode. Our approach is also power aware in the same sense (since we also affect the duration of sensors' operation), but additionally we also control the transmission range (and thus the power dissipation).

Another centralized approach ([8]), is "sensor specific", in the sense it uses some smart powerful sensors that have high processing abilities. In particular, this algorithm assumes that each node is aware of its absolute location via a GPS or a relative location. The sensors must be capable of estimating the distance of the target from the sensor readings.

Distributed Approaches. As opposed to centralized processing, in a distributed model sensor networks distribute the computation among sensor nodes. Each sensor unit acquires local, partial, and relatively coarse information from its environment. The network then collaboratively determines a fairly precise estimate based on its coverage and multiplicity of sensing modalities. Several such distributed approaches have been proposed. Although we are not comparing with them, we shortly discuss some of them, for completeness.

In [12], a cluster-based distributed tracking scheme is provided. The sensor network is logically partitioned into local collaborative groups. Each group is responsible for providing information on a target and tracking it. Sensors that can jointly provide the most accurate information on a target (in this case, those that are nearest to the target) form a group. As the target moves, the local region must move with it; hence groups are dynamic with nodes dropping out and others joining in. It is clear that time synchronization is a major prerequisite for this approach to work. Furthermore, this algorithm works well for merging multiple tracks corresponding to the same target. However, if two targets come very close to each other, then the mechanism described will be unable to distinguish between them.

Another nice distributed approach is the dynamic convoy tree-based collaboration (DCTC) framework that has been proposed in [17]. The convoy tree includes sensor nodes around the detected target, and the tree progressively adapts itself to add more nodes and prune some nodes as the target moves. In particular, as the target moves, some nodes lying upstream of the moving path will drift farther away from the target and will be pruned from the convoy tree. On the other hand, some free nodes lying on the projected moving path will soon need to join the collaborative tracking. As the tree further adapts itself according to the movement of the target, the root will be too far away from the target, which introduces the need to relocate a new root and reconfigure the convoy tree accordingly. If the moving targets trail is known a priori and each node has knowledge about the global network topology, it is possible for the tracking nodes to agree on an optimal convoy tree structure; these are at the same time the main weaknesses of the protocol, since in many real scenaria such assumptions are unrealistic.

The interested reader is encouraged to refer to [18], the nice book by F. Zhao and L. Guibas, that even presents the tracking problem as a "canonical" problem for wireless sensor networks. Also, several tracking approaches are presented in [16].

2   The Model and Its Combinatorial Abstraction

2.1   Sensors and Sensor Network

We abstract the most important technological specifications of existing wireless sensor systems. Each sensor in our model is a fully-autonomous computing and communication device, equipped with a set of monitors (e.g. sensors for temperature, humidity etc.) and characterized mainly by its available power supply (battery) and the energy cost of (1) sensing (i.e. receiving and processing) (2) data sending. The *sensing range* $R_1(t)$ is of particular importance here. We also assume that the sensor can transmit data (to nearly devices) within a range $R_2(t)$. Both these ranges may vary with time. This means that the power spending by the sensor can be set at various different levels. We also assume some law of energy consumption that is range-dependent (e.g. the order of the energy spent is quadratic in the transmitting distance; depending on environmental conditions and their harshness the exponent can be larger than 2).

Let n be the total number of *available* sensor devices for deployment. Let S be the set of *possible* deployment positions (i.e. the union of 3D curves as we described earlier). Let T be a period of time. For each sensor σ of the n available sensors, a placement act, A_σ, is a decision (i) either not to deploy σ or (ii) to deploy σ in an initial position $\overrightarrow{p}$ in S, for a period $T_\sigma \subseteq T$, with a pre-specified pattern of $R_1(t), R_2(t)$ ($t \in T_\sigma$) and a possible trajectory of σ moving in S for all t in T_σ. All the placement acts, together, form a sensor network N. We assume here that N is capable of (somehow) reporting the sensing of local events (i.e. tracking events) to some set of sinks T_N.

2.2   A Model for Targets

In contrast to models that allow only a single moving target, we allow *multiple* targets. We assume that *the initial positions of all targets are arbitrary*. Also, we assume that *all targets arise in arbitrary times* during the network operation.

With respect to target mobility, we assume that each target follows an arbitrary path in S which is however continuous i.e. we disallow the target to instantaneously "jump" to another location. Furthermore, *we do not limit the movement speed of targets*; we only assume that when a target enters the sensing area of an (awake) sensor, it does not manage to leave this area before being sensed. This limit on motion speed is rather trivial, since sensing speed is very high i.e. practically the time needed for a target to be sensed is very close to zero.

2.3   The Combinatorial Model

Assume a given complicated 3D domain, S, with obstacles that disallow signal transmissions (e.g. a set of corridors in buildings or different shape obstacles in a mountain). By "complicated" we mean that the domain can be represented by *arbitrary* three-dimensional curves (see also the problem description subsection), i.e. the only modeling restriction on the curves is their continuity. We further assume that the domain can be represented by the union of λ routes $R_1, \ldots, R_\lambda$ (each can be realized by e.g. a moving robot or air-vessel). We are also given a period T of time.

We wish to equip each route with sensors (of varying capabilities e.g. varying transmission range and operation times) so that any moving object in the domain at any time t in T can be "seen" by at least three sensors (and, thus, its instantaneous position can be found by triangulation). If we can manage this, we can monitor the motion of any moving object within S during T.

A trivial, but very costly, solution is to equip *each route* with sensors (in various points of the route) each operating during the whole T and being able to track any motion in a part of the route. However, one could *exploit overlaps of routes at certain places*, sensors that can monitor pieces of several routes ("nearby") during their operation, or even sensors on top of certain moving objects of our own that wander in the maze of routes.

In order to argue about such economic tracking methods, we view each R_i partitioned into several "route pieces" r_{ij}. Let n_i be the number of the pieces of R_i. We also partition the period T into suitable *intervals* $\tau_1, \ldots, \tau_k$ so that $T = \tau_1 + \cdots + \tau_k$.

We call an *"element"* each pair (r_{ij}, τ_m) , for $1 \leq i \leq \lambda$, $1 \leq j \leq n_i$ and $1 \leq m \leq k$. There are $n = \left(\sum_{i=1}^{\lambda} n_i \right) \cdot k$ such elements.

We can then describe sensor placements (that work for a certain duration each) as relations (sets) between those elements. For example: (a) a sensor placed at r_{i5} can also "see" r_{j7}, r_{k30}. This sensors can operate for 3 intervals. If we start it at τ_1 then the element (r_{i5}, τ_1) "covers" the elements (r_{i5}, τ_2), (r_{i5}, τ_3) but also (r_{j7}, τ_1), (r_{j7}, τ_2), (r_{j7}, τ_3) and (r_{k30}, τ_1) (r_{k30}, τ_2) (r_{k30}, τ_3). (b) A sensor is attached to a moving object that moves from r_{11} to r_{12} to r_{13} and then r_{34}, r_{35}, r_{36}. The sensor lasts 6 intervals and our moving object starts at τ_5. Then, element (r_{11}, τ_5) "covers" elements (r_{12}, τ_6), (r_{13}, τ_7) and also (r_{34}, τ_8), (r_{35}, τ_9), (r_{36}, τ_{10}) (and itself, of course).

In general, each placement of a certain sensor activated at a certain time and operating for a certain time, corresponds to a *set of "covered" elements*. Each such "set" has *a certain cost* e.g. it is more expensive if the sensor's battery is such that the sensor lasts for a long time. Also it is more expensive if its sensing range is large and can "see" more route pieces.

Definition 1. A redundant monitoring design (RMD) D is a set of possible choices of $q \leq n$ sensors $\sigma_1, \ldots, \sigma_q$, each with a placement act $A(\sigma_i)$ of the form "deploy" and the associated cost of the placement act.

Each RMD results in a family of sets $\Sigma_1, \ldots, \Sigma_q$, each having a cost $c(\Sigma_i) > 0$ and each being a subset of our universal set of elements $U = \{e : e \text{ is a pair} (r_{ij}, \tau_m)\}$.

For *feasibility* of the RMD we can require:

(a) that the union of all Σ_i is U.
(b) that each e in U belongs to at least 3 Σ_i sets initially.

However, this is not necessary for our method, since the method itself will discover an infeasible RMD.

Definition 2. Given are an instance of an RMD of sets $\Sigma_1, \ldots, \Sigma_q$ on the universe U of elements related to the domain S and the period T, and also a cost $c(\Sigma_i) \geq 0$ for each Σ_i. Then, an *optimal* final monitoring decision (Optimal FMD) is a sub-collection, F, of sets $\{\Sigma_{t_1}, \ldots, \Sigma_{t_{q'}}\}$ (where $q' \leq q$ and each $t_i \in \{1, 2, \ldots, q\}$), whose total cost $c(F) = \sum_{j=1}^{q'} c(\Sigma_{t_j})$ is minimum, and such that (i) $\cup_i \sum_{t_i} = U$ (ii) each e in U belongs to at least 3 sets in the sub-collection F.

We note that we can construct several RMDs for each run of our Algorithm, get a close to optimal solution (FMD) for each and select the best among them.

3   A Way to Compute Near Optimal FMDs

From the above formulation, the Optimal FMD problem is actually the following "AT-LEAST-3-SET-COVER" problem:

AT-LEAST-3-SET-COVER ($\geq 3SC$): Given $D = (\Sigma_1, \ldots, \Sigma_q)$ where each $\Sigma_i \subseteq U$ (and the union of all Σ_i is U) and given the costs $c(\Sigma_i) \geq 0$, select a minimum total cost sub-collection F of D so that each element e in U belongs to (is "covered" by) at least 3 sets in F.

Important note: Note that geometry and geometric covers can not help here because the domain since S is a complicated 3D domain that can be highly irregular; also, the timing parts of the elements escape the Cartesian geometry; finally, the cost of each Σ_i is a complicated function of placement decisions of sensors of various capabilities.

The problem of $\geq 3SC$ is NP-hard. This is so, since the usual min-cost SET-COVER problem can be reduced to it by adding to any instance of SET-COVER two sets of zero cost, each covering all elements.

We now describe a formulation of the problem as an integer linear program and give an approximate solution based on randomized rounding. Note that our method can be extended to low cost "SET-COVER by a at least l sets" (let us denote this problem as $\geq l$-SET-COVER if the redundancy of $l > 3$ is needed to make the *redundant monitoring decision* D easier to construct and fault-tolerant.

We remind the reader of the computational complexity of set covering ptoblems: Lund and Yannakakis ([13]) showed in 1994 that set cover cannot be approximated in polynomial time to within a logarihmic factor, unless NP has quasi-polynomial time algorithms. Feige ([6]) improved their inapproximability lower bound, under the same assumptions, giving a slightly better bound that essentially matches the approximation ratio achieved by the greedy algorithm. Alon, Moshkovitz, and Safra established in [1] a larger lower bound, under the weaker assumption that P $\neq$ NP. The k-set cover problem is a variant in which every set is of size bounded by k. While k-set cover problem can be solved in polynomial time (via matchings) for $k = 2$, it is NP-complete and even MAX SNP-hard for $k \geq 3$. Greedy algorithms in that case achieve a approximation ratio of $\ln k + \Theta(1)$. Hardness results in [6] show that it is not approximable within $(1 - \epsilon) \ln n$, under strong complexity-theoretic evidence. For 3-set cover, besides linear programming techniques (fractional covers), also local and "semi-local" approximation techniques have been used e.g. in [9,10,5].

3.1   The Randomized Rounding Method

For set Σ let $x(\Sigma)$ be 1 if Σ is selected and 0 else. We want to

$$\text{minimize} \sum_{\Sigma \text{ in } D} x(\Sigma) \cdot c(\Sigma)$$

subject to
(1) $x(\Sigma) \in \{0,1\}$
(2) Let $E(e)$ be the collection of all Σ containing element e. Then, for each element e,

$$\sum_{\Sigma \text{ in } E_e} x(\Sigma) \geq 3$$

Let IP1 be the above integer linear program.

We relax the above integer program to the following linear program (LP1):

$$\text{minimize} \sum_{\Sigma \text{ in } D} x(\Sigma) c(\Sigma)$$

given that
(1)$\forall \Sigma, 0 \leq x(\Sigma) \leq 1$
(2) $\forall e, \sum_{\Sigma \text{ in } E(e)} x(\Sigma) \geq 3$

Let $\{p(\Sigma)\}$ be the optimal solution to the above (i.e. $x(\Sigma) = p(\Sigma)$ get the minimum). We can find $\{p(\Sigma)\}$ in polynomial time in the size n of the redundant monitoring decision D.

We then form a subcollection of sets as follows:

Initially $C = \emptyset$
Experiment E.
For each Σ in D, put Σ into the sub-collection C with probability $p(\Sigma)$, independently of the others.

The experiment E above outputs a sub-collection C. Clearly,

$$E(cost(C)) = \sum_{\Sigma \text{ in } D} c(\Sigma) \Pr\{\Sigma \text{ is picked}\} = \sum_{\Sigma \text{ in } D} c(\Sigma) p(\Sigma) = OPT_f$$

where OPT_f is the optimal value of the linear program LP1. The found collection C has a very nice expected cost (even better than what one can achieve in our original *integer* problem) but we have to examine feasibility.

Since we want to get a C where each e in U belongs to at least 3 sets of C, we must examine whether the (random) C obtained has this property.

Definition 3. *Let α be an element in U and C obtained by the experiment E. We denote by $p(\alpha, C)$ the probability that α belongs to at least 3 sets of C.*

Let w.l.o.g. $\Sigma_1, \ldots, \Sigma_\lambda$ be the sets of our RMD containing element α. Here, we must have $\lambda \geq 3$, else LP1 will report infeasibility. W.l.o.g. denote by p_i the probability $p(\Sigma_i)$ obtained via LP1, i.e. that Σ_i is chosen to be in C. Assuming that LP1 is feasible we get:

$$\gamma = p_1 + \cdots + p_\lambda \geq 3 \tag{1}$$

Let A_3, N_0, N_1, N_2 be the events:
$A_3 = $ "α is covered by at least 3 sets in C"
$N_0 = $ "α does not belong to any set in C"
$N_1 = $ "α belongs to exactly one set in C"
$N_2 = $ "α belongs to exactly two sets in C"
Now,

$$p(\alpha, C) = \Pr\{A_3\} = 1 - \Pr\{N_0\} - \Pr\{N_1\} - \Pr\{N_2\} \tag{2}$$

We now estimate $\Pr\{N_i\}$, for $i = 0, 1, 2$:
We will repeatedly use the following fact:

Fact (*). If the numbers $x_1, \ldots, x_\lambda$ are each in $[0, 1]$ and $x_1 + \cdots + x_\lambda \geq \gamma$, then $(1 - x_1) \cdots (1 - x_\lambda) \leq \left(1 - \frac{\gamma}{\lambda}\right)^\lambda \leq e^{-\gamma}$ (it is assumed that $\gamma \leq \lambda$).

Fact (*) can be proved via an easy induction, or via the fact that the arithmetic mean is bigger or equal than the geometric one.

Proof of Fact (*). Let $y_i = 1 - x_i, i = 1, \ldots, \lambda$. Then, $\sum y_i = \lambda - \sum x_i \leq \lambda - \gamma$. So,

$$\frac{\sum y_i}{\lambda} \leq 1 - \frac{\gamma}{\lambda} \tag{3}$$

But, $\frac{\sum y_i}{\lambda} \geq \sqrt[\lambda]{\pi y_i}$ (arithmetic mean vs geometric mean), so:

$$\frac{\sum y_i}{\lambda} \geq \sqrt[\lambda]{(1 - x_1) \cdots (1 - x_\lambda)}$$

So, by Equation (3), $\sqrt[\lambda]{(1 - x_1) \cdots (1 - x_\lambda)} \leq 1 - \frac{\gamma}{\lambda}$. So, $(1 - x_1) \cdots (1 - x_\lambda) \leq \left(1 - \frac{\gamma}{\lambda}\right)^\lambda$ ◇

We now have:
(a) $\Pr\{N_0\} = (1 - p_1) \cdots (1 - p_\lambda)$. By the Fact (*), then

$$\Pr\{N_0\} \leq \left(1 - \frac{\gamma}{\lambda}\right)^\lambda \leq \exp(-\gamma).$$

Also, by inclusion-exclusion,

$$\Pr\{N_1\} = p_1 \cdot \Pr\{\text{no cover of } \alpha \text{ in the trials of } \Sigma_2, \ldots, \Sigma_\lambda\}$$

$$+(1-p_1)\Pr\{1 \text{ cover of } \alpha \text{ in the trials of } \Sigma_2, \ldots, \Sigma_\lambda\} \leq p_1 \cdot \frac{1}{e^{\gamma - p_1}} + (1-p_1)\Pr\{N_1\}$$

(because $p_2 + \cdots + p_\lambda = \gamma - p_1$ so $(1-p_2)\cdots(1-p_\lambda) \leq \left(1 - \frac{\gamma - p_1}{\lambda}\right)^\lambda = \exp(-(\gamma - p_1))$ and also because $\Pr\{1 \text{ success in the trials of } \Sigma_2, \ldots, \Sigma_\lambda\} \leq \Pr\{N_1\}$).
So,

$$\Pr\{N_1\} \leq p_1 \cdot \frac{1}{e^{\gamma - p_1}} + (1-p_1)\Pr\{N_1\}$$

so

$$\Pr\{N_1\} \leq \frac{1}{e^{\gamma - p_1}} \leq \frac{1}{e^{\gamma - 1}}$$

Similarly,

$$\Pr\{N_2\} = p_1 \cdot \Pr\{1 \text{ cover of } \alpha \text{ in the trials of } \Sigma_2, \ldots, \Sigma_\lambda\}$$

$$+(1-p_1)\Pr\{2 \text{ covers of } \alpha \text{ in the trials of } \Sigma_2, \ldots, \Sigma_\lambda\} \leq p_1 \cdot \frac{1}{e^{\gamma - p_1}} + (1-p_1)\Pr\{N_2\}$$

also, so $\Pr\{N_2\} \leq \frac{1}{e^{\gamma - 1}}$.
Thus,

$$\Pr\{A_3\} = p(a, c) \geq 1 - \frac{1}{e^\gamma} - \frac{2}{e^{\gamma - 1}}$$

and $\gamma \geq 3$. So, $\Pr\{A_3\} = p(a, c) \geq 1 - \frac{1}{e^3} - \frac{2}{e^2} > \frac{1}{2}$. Let $\xi = 1 - \frac{1}{e^3} - \frac{2}{e^2}$. Hence, we get the following:

Theorem 1

$$\Pr\{\alpha \text{ is covered by at least 3 sets in } C\} = p(\alpha, c) \geq 1 - \frac{1}{e^3} - \frac{2}{e^2} = \xi$$

We now repeat the experiment E (with the same $p(\Sigma_i)$) to get $r = c\log n$ such collections $C_1, C_2, \ldots, C_r$.

Let $V = C_1 \cup C_2 \cup \ldots \cup C_r$. By independence of the repeated experiments, if the event AV_3 is :
$AV_3 =$ "element α is not covered by at least 3 sets in V" then

$$\Pr\{AV_3\} \leq (1 - \xi)^{c\log n}$$

We can always choose c so that $(1 - \xi)^c < 1/4$. Then,

$$\Pr\{AV_3\} \leq \left(\frac{1}{4}\right)^{\log n} \leq \frac{1}{n^2}$$

Thus the probability that there is an element in U not covered by at least 3 sets of V is bounded by above by $n \cdot n^{-2} = 1/n$.
So, we get the following:

Theorem 2. The collection V obtained satisfies: (i)

$$\Pr\{V \text{ covers each element by at least 3 sets}\} \leq 1 - \frac{1}{n}$$

and (ii)

$$E(cost(V)) \leq c \log n \ OPT_f \leq c \log n \ OPT$$

where OPT is the cost of the *optimal* FMD.

Let $\rho = c \log n$. Note that by the Markov inequality it is

$$\Pr\{cost(V) < 2\rho \, OPT_f\} \geq \frac{1}{2}$$

Thus, the probability that V is valid and has cost less than $2\rho OPT_f$ is at least

$$1 - \left(\frac{1}{n} + \frac{1}{2}\right) = \frac{1}{2} - \frac{1}{n}$$

We can then repeat the whole process an expected number of at most 2 times and get a V which is verified to be an almost optimal and valid FMD. Note that we have also shown:

Theorem 3. The problem $\geq 3SC$ can be approximated in polynomial expected time with an approximation ratio $\Theta(\log n)$.

4   The Triangulation Issue

The solution to $\geq 3SC$ of the last section selects a close to optimal FMD. Thus, it also specifies the initial positions of the associated (selected) sensors. However, no guarantee is provided that for any element e, the 3 elements "covering" e actually form a triangle (e.g. they may be on a line). We propose to handle this via a "post-processing" step as follows:

For each $e = (r_{ij}, \tau_m)$ let $\sigma_1(e), \sigma_2(e), \sigma_3(e)$ be the 3 sensors covering e (i.e. covering r_{ij} at τ_m). We now modify their placement act by perturbing the positions p_i^* of $\sigma_i(e)$ at τ_m by a random, small perturbation of center their p_i^* and radius $\epsilon > 0$, small enough so that the same sets are covered by them. We perform the perturbation act only for those e for which the $\sigma_i(e)$ are not forming a triangle.

At the end of the post-processing step, each e in U is covered by (at least) 3 sensors, whose positions (during τ_m) form a triangle with high probability. We can repeat the perturbation until the triangle is indeed formed.

This post-processing step can always be done, provided that S gives an infinitesimal free space around each of its points. This is safe to assume for any application. We note that our solution works for static sensors, and moving sensors whose motion is controlled by us, while it can not be applied to the case of sensors moving on their own.

5   Alternative Collaborative Processing Methods for Our Approach

Our method is based on an easy to get RMD which is then "cleared" via the randomized rounding technique to get a low cost final monitoring decision. Its result is a selection of sensor placement acts, guaranteed to monitor each point $\vec{p}$ in S by at least 3 sensors for any time in the period T. When our localization method is combined with ability to distinguish all observed objects (e.g. by using unique IDs) then tracking of objects can be performed.

Clearly our method must be complemented by a way to report target positions and their associated times to some central facility. If the reporting delay is comparable to the speed of the target then the central facility can reconstruct the targets motion in real time. To this end, Delay and Disruption Tolerant Networking (DTN, see e.g. [11]) approaches can be useful, to improve network communication when connectivity is periodic, intermittent or prone to disruptions and when multiple heterogeneous underlying networks may need to be utilized to effect data transfers.

The sensor network may handle the localization reports (i.e. reports on target position at a certain time) distributedly. This reduces communication cost but the central facility must have a way to compare those reports into a consistent information about the trajectory of the target. The issues of local clocks and their synchronization is crucial for this and we do not solve it here. In fact, we view our approach as a building block for full tracking approaches.

6   Conclusions

To design low cost sensor networks able to efficiently localize and track multiple moving targets, we try to take advantage of possible coverage ovelaps over space and time, by introducing an abstract, combinatorial model that captures such overlaps. Under this model, we abstract the localization and tracking problem by a combinatorial problem of set covering (by at least three sets, to ensure localization of any point in the network area, via triangulation). We then provide an efficient approximate method for sensor placement and operation, that with high probability and in polynomial expected time achieves monitoring decisions with a $\Theta(\log n)$ approximation ratio to the optimal solution.

We can start with redundant monitoring decisions D giving least frequency of covering an element which is very high. This allows a high flexibility in possible places, timings and motions of sensors to initially "over-cover" the domain. Then, we can run our algorithm to choose an economic implementation. Note that our solution is of low cost and also achieves continuous monitoring of any moving object in the period T. By extending our techniques, we can prove that the problem $\geq \lambda$-SET-COVER (λ any constant) has a (polynomial time) approximation ratio of $\Theta(\log n)$.

References

1. Alon, N., Moshkovitz, D., Safra, M.: Algorithmic construction of sets for k-restrictions. ACM Transactions on Algorithms (TALG) 2(2), 153–177 (2006)
2. Aspnes, J., Goldberg, D., Yang, Y.: On the computational complexity of sensor network localization. In: Nikoletseas, S.E., Rolim, J.D.P. (eds.) ALGOSENSORS 2004. LNCS, vol. 3121, pp. 32–44. Springer, Heidelberg (2004)
3. Chakrabarty, K., Iyengar, S.S., Qi, H., Cho, E.: Grid coverage for surveillance and target location in distributed sensor networks. IEEE Trans. Comput. 51(12) (2002)
4. Dhilon, S.S., Chakrabarty, K., Iyengar, S.S.: Sensor placement for grid coverage under imprecise detections. In: FUSION 2002. Proc. 5th Int. Conf. Information Fusion, Annapolis, MD, pp. 1–10 (July 2002)
5. Duh, R.-c., Fuerer, M.: Approximation of k -Set Cover by Semi-Local Optimization. In: ACM STOC (1997)
6. Feige, U.: A Threshold of ln n for Approximating Set Cover. Journal of the ACM (JACM) 45(4), 634–652 (1998)
7. Gui, C., Mohapatra, P.: Power conservation and quality of surveillance in target tracking sensor networks. In: Proc. ACM MobiCom Conference (2004)
8. Gupta, R., Das, S.R.: Tracking moving targets in a smart sensor network. In: Proc. VTC Symp. (2003)
9. Halldorsson, M.M.: Approximating Discrete Collections via Local Improvements. In: ACM SODA (1995)
10. Halldorsson, M.M.: Approximating k-Set Cover and Complementary Graph Coloring. In: Cunningham, W.H., Queyranne, M., McCormick, S.T. (eds.) Integer Programming and Combinatorial Optimization. LNCS, vol. 1084, Springer, Heidelberg (1996)
11. Jain, S., Demmer, M., Patra, R., Fall, K.: Using Redundancy to Cope with Failures in a Delay Tolerant Network. In: Proc. SIGCOMM 2005 (2005)
12. Liu, J., Liu, J., Reich, J., Cheung, P., Zhao, F.: Distributed group management for track initiation and maintenance in target localization applications. In: Zhao, F., Guibas, L.J. (eds.) IPSN 2003. LNCS, vol. 2634, Springer, Heidelberg (2003)
13. Lund, C., Yannakakis, M.: On the hardness of approximating minimization problems. Journal of the ACM (JACM) 41(5), 960–981 (1994)
14. Poduri, S., Sukhatme, G.S.: Constrained coverage in mobile sensor networks. In: ICRA 2004. Proc. IEEE Int. Conf. Robotics and Automation, New Orleans, LA, April-May 2004, pp. 40–50 (2004)
15. Rao, N.S.V.: Computational complexity issues in operative diagnosis of graph based systems. IEEE Trans. Comput. 42(4) (April 1993)
16. Shorey, R., Ananda, A., Chan, M.C., Ooi, W.T.: Mobile, Wireless, and Sensor Networks: Technology, Applications, and Future Directions. Wiley, Chichester (2006)
17. Zhang, W., Cao, G.: Optimizing tree reconfiguration for mobile target tracking in sensor networks. In: Proc. IEEE InfoCom (2004)
18. Zhao, F., Guibas, L.: Wireless Sensor Networks: An Information Processing Approach. Morgan Kaufmann, San Francisco (2004)
19. Zou, Y., Chakrabarty, K.: Sensor deployment and target localization based on virtual forces. In: Proc. IEEE InfoCom, San Francisco, CA, pp. 1293–1303 (April 2003)

Counting Targets with Mobile Sensors in an Unknown Environment

Beat Gfeller[1], Matúš Mihalák[1], Subhash Suri[2,*],
Elias Vicari[1,**], and Peter Widmayer[1,**]

[1] Department of Computer Science, ETH Zurich, Zurich, Switzerland
{gfeller,mmihalak,vicariel,widmayer}@inf.ethz.ch
[2] Department of Computer Science, University of California, Santa Barbara, USA
suri@cs.ucsb.edu

Abstract. We consider the problem of counting the number of *indistinguishable* targets using a simple binary sensing model. Our setting includes an unknown number of point targets in a (simply- or multiply-connected) polygonal workspace, and a moving point-robot whose sensory input at any location is a binary vector representing the cyclic order of the polygon vertices and targets visible to the robot. In particular, the sensing model provides no coordinates, distance or angle measurements. We investigate this problem under two natural models of environment, friendly and hostile, which differ only in whether the robot can visit the targets or not, and under three different models of motion capability.

In the friendly scenario we show that the robots can count the targets, whereas in the hostile scenario no $(2 - \varepsilon)$-approximation is possible, for any $\varepsilon > 0$. Next we consider two, possibly minimally more powerful robots that can count the targets exactly.

1 The Problem and the Model

Simple, small and inexpensive computational and sensing devices are currently at the forefront of several research areas in computer science. These devices promise to bring computational capabilities into areas where previous approaches (usually consisting of complex and bulky hardware) are not feasible or cost-effective. Such devices are being successfully used in various monitoring systems, military tasks, and other information processing scenarios. Their main advantages are quick and easy deployment, scalability, and cost-effectiveness. However, in order to realize the full potential of these technologies, many new and challenging

* Work done while the author was a visiting professor at the Institute of Theoretical Computer Science, ETH, Zurich. The author wishes to acknowledge the support provided by the National Science Foundation under grants CNS-0626954 and CCF-0514738.
** Work partially supported by the National Competence Center in Research on Mobile Information and Communication Systems NCCR-MICS, a center supported by the Swiss National Science Foundation under grant number 5005 − 67322.

M. Kutyłowski et al. (Eds.): ALGOSENSORS 2007, LNCS 4837, pp. 32–45, 2008.

research problems must be solved, because the classical schemes designed for centralized and desktop computational hardware are inapplicable to the lightweight and distributed computational model of sensor nodes. The inherent limitations of the systems based on these simple devices have inspired the research community to consider the computation with a minimalistic view of hardware complexity, sensing and processing, energy supply, etc.

In this paper we use such a minimalistic approach in the area of mobile sensors – simple robots. We consider and define robots of unsophisticated sensing and mobile capabilities and investigate their computational power on an elementary yet natural problem of counting objects of interest in the robots' environment. We model the environment by a polygon P (simply or multiply-connected) in the plane and the objects of interest, namely, *targets* are modeled as a set of points inside P.

We assume that the robot is a (moving) point, equipped with a simple camera that can *sense* just the combinatorial features of the surrounding. In particular, the robot can see a vertex of P or a target, can distinguish a target from a vertex, but the vertices and the targets are otherwise indistinguishable, i.e., all vertices are visually identical and all targets are visually identical. It is only the cyclic order in which the robot sees the features that distinguishes them from each other. We assume that the ordering is always consistent, which we take, without loss of generality, to be counterclockwise. We model such a discrete vision by a *point identification vector* (piv), which is a binary vector defined by the cyclically ordered list of targets and polygon vertices that are visible from the current robot's position, where each bit indicates whether the corresponding point is a target (value 1) or a vertex of the polygon (value 0). Sitting at a vertex of P, we assume that the cyclic order of the visible points (vertices and targets) starts with the neighboring vertex, i.e., the first component of the piv always represents the neighboring vertex (and therefore has value 0). For the robot located on a target, we make no assumption about the first component of the robot's piv – it is chosen by an adversary.

Moreover, the robot can see the edges of the polygon. This is modeled by a *combinatorial visibility vector* (cvv), a binary vector of length k whose i-th bit encodes whether there is an edge between vertex $i - 1$ and vertex i of the k vertices visible from the robot's position. See Fig. 1 for illustration.

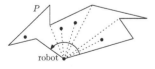

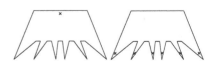

Fig. 1. An illustration of a point identification vector (piv) and a combinatorial visibility vector (cvv) in polygon P (with 4 targets); piv is $(0, 1, 0, 0, 1, 1, 0, 0)$ and cvv is $(1, 1, 0, 1, 0, 1)$

Fig. 2. If a robot only senses the number of targets then the number of targets cannot be approximated within $o(n)$

The robot has no other sensing capability, and in particular has no information about distances, angles, or world coordinates. This also motivates our simplistic model of the robot's movement. Roughly speaking, the robot can pick a direction based on its sensing system and move in that direction until the environment prevents the robot to go any further. The direction of a robot's movement is the direction to one of the visible points (a vertex of P or a target) in the robot's piv, and the robot stops when it reaches that point. The robot can sense the environment only when it is not moving.

The robots model simple and small mobile sensors, which possess a low-resolution camera and limited computational power, which allows the robots to perform only simple image processing tasks, such as finding areas of substantial light changes. The limited power and equipment prevents learning anything more, like distances, angles, etc., and thus a binary sensing reflects appropriately what robots sense. The robots shall explore an unknown environment, which is physically bounded, such as buildings (with walls), or streets of a city. Thus we naturally arrive at the model with a polygon P and discrete sensing via piv and cvv.

Due to these unsophisticated vision and motion primitives, seemingly easy tasks become difficult in this model. For instance, a robot sitting at a vertex u can specify a visible vertex v by its index in the cvv of u. However, if the robot moves from u to v, it is not possible in general to recover the position of u with respect to v. A way to circumvent this issue is to mark u with a *pebble* before moving to v. A pebble is visually distinguishable from vertices and targets. If no other pebbles are visible from v, the position of u can be recovered. For a detailed discussion of the implications of this minimalistic model, see [1].

We are interested in how the robots can solve various environment exploration tasks and what limitations are implied by our simplistic assumptions on robots' capabilities. In this paper we consider the problem of determining the number of targets in an unknown polygon P. Throughout this paper we refer to this problem as the *counting problem*. By n we denote the number of vertices of P and by m the number of targets therein. For simplicity we assume that the targets and polygon vertices are in a general position, i.e., no three points are collinear. In this paper we consider two different scenarios to model two basic classes of applications. In the *friendly* environment, the robot is allowed to walk to any target. In the *hostile* environment, the robot is not allowed to walk to targets. This scenario models the situation where a target represents an unsafe entity and coming into an imminent closeness to targets is dangerous.

For the friendly scenario we show that a single robot with two pebbles (markers) can count the number of targets in any polygon P. In contrast, we show that in the hostile scenario, the robots cannot count the targets in general. Thus, requiring the robots to count targets only from afar is a more complicated problem, and we must endow the robots with some additional capabilities. Surprisingly, we show that these additional capabilities are quite minor, yet subtle. In fact, we consider two possible models, and show their implications on our problem. We consider robots that can walk along edge or diagonal extensions, i.e., if a robot picks a visible vertex u as the direction of its walk, the robot can continue its walk in the same direction after it reaches u, if there is no polygonal edge to prevent it. In the second model we consider one additional global direction (think of "north") in which the robot can walk from any vertex of P. In both models the robots can solve the counting problem.

We are interested in deterministic algorithms and their worst-case analysis, which we express in terms of the number of steps (movements) of robots and in the amount of used memory. We work with word-memory units, where one word of memory has $\Theta(\log(\max\{m, n\}))$ bits. We are also interested in approximation algorithms, i.e., in algorithms that deliver a (provably good) estimate on the number of targets. Further, we look for estimates that are never smaller than the actual number of targets. We say that an algorithm is a ρ-approximation for the counting problem if for the setting with m targets, $m \in \mathbb{N}$, the algorithm estimates the number of targets by z, for which $m \leq z \leq \rho \cdot m$.

To demonstrate the notion of approximation and to justify our sensing model we illustrate that for the following weaker sensing model no non-trivial approximation exists: consider the sensing of the vertices in the same way as we defined before, but consider the sensing of the targets only by their presence, i.e., not interleaved with the vertices. Thus, the only information the robot gets is the number of visible targets (but not their ordering within the vertices of P). Consider Fig. 2. It depicts two different scenarios, one scenario with $m = 1$ target and the second scenario with $m = n/3$ targets. In both scenarios the robot senses from every vertex exactly one target and therefore cannot distinguish the scenarios. Hence, for this simple sensing model no approximation algorithm can guarantee a ratio better than $n/3$.

Related Work. Suri et al. [1] considered simple robots with combinatorial sensing of the environment and investigated some elementary questions of what information about the topology of the environment can be deduced by simple robots. In our paper, we consider the same robots, but enlarge the complexity of the environment by adding the targets into the environment. Although the robots are strongly limited in capabilities, it is shown in [1] that the robots can solve non-trivial tasks. A robot cannot decide whether a vertex is convex, but can decide whether the polygon is convex. Also, a robot cannot decide which is the outer boundary of a multiply-connected polygon P, although it can discover and count all the boundary components in P. Furthermore, a robot with one pebble can build a mental map of the vertices of any (simply- or multiply-connected) polygon P in $O(n^3)$ steps and with $O(n)$ memory, which allows the robot to navigate from any vertex i to any vertex j. The navigation result is an important building block in our paper. Further, the paper shows that the robot can compute a triangulation of P and solve a distributed version of the famous Art Gallery Problem [2] with $\lfloor n/3 \rfloor$ guarding positions.

Combinatorial geometric reasoning is used in many motion planning and exploration tasks in robotics [3,4]. Minimalistic models of robots has been previously studied in [5,6,7,8]. However, the nature of problems investigated in our paper does not seem to be addressed in the past. The aforementioned papers deal with different problems such as navigation and pursuit evasion [6,7,8], and not with recognition of important points (targets) in the environment. Learning about the geometrical nature of the environment is the problem studied in [5], where the environment is not a polygon, and it contains *labeled* features, which allows sensors to distinguish these *landmarks*.

2   The Friendly Environment

In this section we show that in a friendly environment a robot with two pebbles can count the targets in any simply or multiply connected polygon.

We consider simply-connected polygons first. In the beginning the robot counts n, the number of vertices of the polygon: the robot leaves a pebble on the starting vertex and walks around the polygon's boundary, counting the vertices until it returns to the pebble. Let $1, 2, \ldots, n$ denote the vertices of the polygon, ordered in the counterclockwise direction, starting at the robot's initial position.

The idea of the algorithm is to go to every vertex i, $i = 1, 2, \ldots, n$, and count the targets that are visible from i and that are not visible from any vertex j, $j < i$. We call these targets *newly visible* at vertex i. Thus, the robot can go through vertices $i = 1, 2, \ldots, n$ and sum up all newly visible targets. Clearly, no target will be counted twice, and therefore the resulting sum is the total number of targets.

We now describe how the robot can identify whether a target is newly visible. Being at vertex i, the robot wants to identify whether a k-th target in its visibility vector is newly visible. The robot goes to the target, leaves a pebble there, and checks for every vertex $j < i$, whether the pebble is visible from j. The navigation from the target back to the vertex i can be done by leaving the second pebble at i and checking the position of i in the visibility vector of the target. Obviously, the target is newly visible if and only if the pebble is not visible from any vertex j, $j < i$. Overall, the robot needs two pebbles and a constant number of memory words (to remember the number of vertices, the current position i, the position j and the position k of the considered target at i, and to mark the newly visible targets in the visibility vector of vertex i). Hence, in $2i$ steps we can check whether a target visible from the i-th vertex is newly visible. To check all targets at position i we need at most $2mi$ steps. Thus, the robot needs $O(mn^2)$ steps to count the targets in P.

If the time is crucial, one can achieve a $O(mn)$ number of steps at the expense of used memory. For each vertex i the robot maintains the piv with the additional information stating whether a given target is newly visible. In the beginning, every target in the piv is marked as newly visible. Then for every vertex i the robot marks each newly visible target with a pebble and walks around the boundary towards vertex n and at every vertex j, if the robot sees the pebble, it marks the corresponding bit in the bit array of vertex j as not newly visible. Thus, the robot walks m times around the boundary (for each target it walks exactly once and at most n steps), resulting into $O(mn)$ steps of the robot. The robot needs $O(nm)$ memory.

Theorem 1. *In the friendly environment a robot with two pebbles can count the targets in a simply-connected polygon in $O(mn^2)$ steps and with $O(1)$ memory, or in $O(mn)$ steps and with $O(nm)$ memory.*

The result can be easily extended to polygons with holes (multiply-connected polygons), if we can navigate through the vertices in a consistent way. In [1], a navigation in an arbitrary multiply connected polygon was demonstrated with a robot with one pebble in a polynomial number of steps and with polynomial space. Our robot has all the capabilities of the robot described there, therefore the robot can first compute the navigation instructions, which are then stored in the robot's memory. Alternatively, we can use an additional, globally distinguishable pebble and perform the vertex navigation on the fly.

Theorem 2. *In the friendly environment a robot with two pebbles can count the targets in any polygon in polynomially many steps and with polynomial memory.*

3   Hostile Environment

After solving the counting problem in the scenario where robots can walk to targets, we consider now the scenario where robots walk only on vertices of P.

3.1   Inapproximability and Approximation

Inapproximability. We show that the counting problem cannot be approximated within a factor $2 - \varepsilon$, for any $\varepsilon > 0$, even if the polygon P is simply-connected. We start with a warm-up example to illustrate the idea. Consider the polygon in Fig. 3. The polygon consists of four spikes attached to the four sides of a rectangle. It depicts two scenarios with a different number of targets. In the first scenario there are 6 targets and in the second scenario there are 4 targets. Considering any vertex of the polygon, the vectors cvv and piv are the same in both scenarios. Hence, the robot cannot distinguish the two scenarios, which shows a lower-bound of $6/4 = 3/2$ for the approximation ratio.

This construction can be extended to a general-sized polygon, where $2k$ spikes are attached to a regular $2k$-gon, using $2k$ and $4k - 2$ targets in two different scenarios, thus giving the desired inapproximability lower-bound of $2 - \varepsilon$.

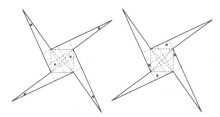

Fig. 3. The counting problem cannot be approximated within a factor $3/2$

Theorem 3. *The counting problem cannot be approximated within a factor* $2-\varepsilon$, *for any* $\varepsilon > 0$, *even in a simply-connected polygon.*

Note that this inapproximability result relies only on the visibility limitations of the robots and not on their limited navigation capabilities.

Proof. We assume n is even, i.e., $n = 2k$. The shape of the polygon is depicted in Fig. 4 and 5. The polygon consists of n *outer* vertices $y_1, y_2, \ldots, y_n$ and n *inner* vertices $x_1, x_2, \ldots, x_n$. It can be viewed as an n-gon, a regular polygon formed by vertices x_i, $i = 1, \ldots, n$, connected on each side x_i, x_{i+1} to a triangular spike x_i, y_i, x_{i+1}. Here and further in the text, the indices are to be understood in a cyclic fashion. The line $y_i x_i$ intersects the segment $x_{i+1} x_{i+2}$ in the middle. Thus, the visibility region of y_i, i.e., the cone of y_i defined by lines y_i, x_i and y_i, x_{i+1}, intersects the visibility regions of vertices y_{i-1} and y_{i+1}, but not the visibility regions of other y_js.

Observe first that a robot at vertex y_i sees only two vertices of P, namely vertex x_i and vertex x_{i+1}. Further, a robot sitting at vertex x_i sees all vertices x_j, $j = 1, 2, \ldots, n$, and vertices y_{i-1} and y_i.

The aim is to place the targets in a way that a robot sitting at vertex y_{2l+1} sees one target (the piv is $(0, 1, 0)$), and a robot sitting at vertex y_{2l} sees 2 targets (the piv is $(0, 1, 1, 0)$). For a robot at vertex x_i, $i = 1, \ldots, n$, we want the robot to see exactly 1 target between each two consecutive vertices of its piv, i.e., we want the piv to be $(0, 1, 0, 1, 0, 1, \ldots, 0, 1, 0)$. Observe that the consecutive vertices of piv at vertex x_i are $y_i, x_{i+1}, \ldots, x_n, x_1, \ldots, x_{i-1}, y_{i-1}$. We show how to achieve such visibility with two different number of targets. First we use only n targets and then we use $2n - 2$ targets.

To place the n targets we proceed as follows. We place one target into each triangle y_i, x_i, x_{i+1}. Observe that the triangle is divided into three parts by the lines y_{i-1}, x_i and y_{i-1}, x_{i-1}. Let us label the parts P_1, P_2 and P_3, starting at a part containing the vertex x_{i+1}. Fig. 4 illustrates the partition. For odd i, we place the target into part P_2. For even i, we place the target into P_1. Observe now that a robot indeed sees one target from every vertex y_{2l+1} and two targets from every vertex y_{2l}. Observe also that any vertex x_j sees exactly one target between two consecutive vertices x_i, x_{i+1}, $i, i+1 \neq j$, because the parts P_1 and P_2 of triangle y_i, x_i, x_{i+1} contain exactly one target and the parts are completely visible from x_j within the segment x_i, x_{i+1}. There is also one target visible in the segment y_j, x_{j+1} and in the segment x_{j-1}, y_j which shows the claim for n targets.

We now use $2n - 2$ targets in P to achieve the same visibility configuration. First, we place one target into every triangle x_i, y_i, x_{i+1} such that the target is visible only from vertices x_i, y_i and x_{i+1}. This can be easily achieved when the target is placed very close to y_i. This leads to piv being $(0, 1, 0)$ at vertices y_i and piv being $(0, 1, 0, 0, \ldots, 0, 0, 1, 0)$ at vertices x_i. The remaining $n - 2$ targets are placed in the following way. For the presentation purposes we label the targets $t_1, \ldots, t_{n-2}$. Each target t_i is placed *close* to vertex x_i and in the cone C_i of x_i defined by the vertices x_{n-1}, x_n. More precisely, by placing t_i close to x_i we mean to place the target t_i into the triangle $T_i := x_{i-1}, x_i, x_{i+1}$. Observe now

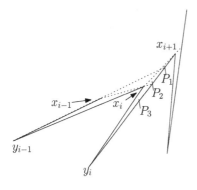

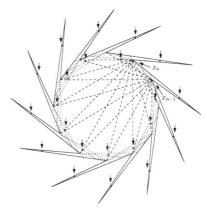

Fig. 4. The partition of the triangle y_i, x_i, x_{i+1} into three parts P_1, P_2 and P_3 by the lines y_{i-1}, x_{i-1} and y_{i-1}, x_i

Fig. 5. A placement of $2n-2$ targets into the polygon P. The arrows indicate the position of the targets.

that for any placement of target t_i into $C_i \cap T_i$ the piv of vertex x_i is as desired, i.e., $(0, 1, 0, 1, 0, \ldots, 0, 1, 0)$. Indeed, for vertex x_i, $i \le n - 2$, the cone C_i contains t_i and thus the target is visible between x_{n-1} and x_n. For every other cone of x_i defined by vertices x_j and x_{j+1}, the target t_j lies in that cone. Also, for vertex x_{n-1} the cone of x_{n-1} defined by vertices x_i and x_{i+1} contains exactly one target, namely t_{i+1}. Similarly, the cone of vertex x_n defined by vertices x_i and x_{i+1} contains exactly one target, namely t_i. To achieve the desired piv from the vertices y_i, we place each target t_i within T_i either to the left or to the right of line y_{i-1}, x_{i-1}. For $i - 1 = 2l$ we place t_i to the right of the line y_{i-1}, x_{i-1}, so that t_i is visible from y_{i-1} (i.e., into the cone of y_{i-1} defined by vertices x_{i-1} and x_i). For $i - 1 = 2l + 1$ we place t_i to the left of line y_{i-1}, x_{i-1}, so that t_i is not visible from y_{i-1}. It is easy to observe that for every vertex y_i, its piv is $(0, 1, 0)$ if $i = 2l + 1$, and $(0, 1, 1, 0)$ if $i = 2l$. A placement of $2n - 2$ targets into the polygon P with $2n$ vertices, where $n = 12$, is depicted in Fig. 5. This ends the proof.   □

Approximation. Since the counting problem cannot be solved optimally, it is natural to look for approximate solutions, i.e., for good estimates of m, the number of targets. Observe first that m is at least the number of targets visible from any vertex of P. Let m_i denote the number of targets that are visible from vertex i. We have $m \ge \max_i m_i$. On the other hand, clearly, $m \le \sum_i m_i$. Since every target is visible from at least three vertices of P (consider a triangulation of P and the vertices of the triangle, in which the target lies), we have $m \le \frac{1}{3} \sum_i m_i$. A robot can compute the sum $z = \sum_i m_i$ with one pebble that allows the robot to navigate through all vertices of P (even with holes [1]). Obviously, reporting $\frac{1}{3} z$ as the estimate for the number of targets yields an $\frac{n}{3}$-approximation. Alternatively, if we denote by k the number of vertices with non-zero m_i, the value z becomes a $\frac{k}{3}$-approximation.

Although the approximation is not sound at first sight (consider a convex polygon with a single target in it), it gives some insight into the complexity of the counting problem. Notice that the derived approximation ratio depends solely on the number of vertices n (or on k, the number of vertices with a view on at least one target) and not on the number of targets. Hence, if m grows in comparison to n or k, the approximation ratio gets better. In other words, the approximation ratio does not grow with the number of targets, but is determined by the structure of the polygon (i.e., by n) and by the way how the targets are placed in this structure (i.e., by k).

A 2-approximation algorithm can be designed under a slightly stronger model [9] (but weaker than the one described in Section 3.3). In this model, the $2 - \varepsilon$ inapproximability result still holds.

3.2   More Power to the Robots

We have seen in the previous subsection that a simple robot cannot count the targets in a simple polygon. We therefore look at possible enhancements of capabilities, which keep the robots as simple as possible and at the same time enable the robots to count the targets. We consider two such enhancements.

In the first one we allow the robots to walk along edge-extensions and diagonal-extensions, i.e., if a robot at vertex v picks a vertex w as the direction of the robot's walk, the robot is allowed to continue walking in the same direction after it reaches w, and it will stop only when it hits the boundary, at a point w'. Fig. 6 illustrates this enhancement. The line vw' is called an *edge-extension* (*diagonal-extension*) if vw is an edge (diagonal) of P. If we do not need to distinguish whether vw' is an edge- or diagonal-extension, we simply say that vw' is an *extension*. If a pebble is placed at w', it is then visible in the same way as a vertex of P, and therefore the robot can go there from any vertex visible to it. w' is then visually distinguishable from the other regular vertices of P, because it is marked with a pebble.

In the second enhancement one additional, global direction is introduced, in which a robot can move. Without loss of generality we assume that it is the direction of a vertical line going through the robot's position. For simplicity of presentation we assume that the polygon does not have vertical edges. On top of that we assume the robot can tell whether a visible point (a vertex or a target) is to the left or right of the vertical line, and whether it is above or below the robot, i.e., above or below the horizontal line going through the robot's position. Such an enhancement can be viewed as a navigation with compass. If a robot walks from a vertex v in the vertical direction we say that it walks along the *vertical extension* of v.

For both enhancements we present algorithms that allow robots to count the number of targets inside the polygon P.

Partition of the Polygon and Counting. The algorithms are based on the idea of partitioning the polygon into triangles and counting the targets in these triangles exactly. To illustrate the idea, consider a partition of P into triangles having their vertices on the boundary of P with the property that every triangle has at least one side on the boundary of P. We call such a triangle a *baseline* triangle, and the edge of the triangle that lies on the boundary of P a *baseline edge*. A partition of a polygon into baseline triangles is called a *baseline triangulation*. We might want to require that the baseline triangles of a baseline triangulation are triangles in the classical sense, i.e., specified only by vertices of P – these are called *baseline vertex triangles*. A triangulation into baseline vertex triangles is called a *Hamiltonian triangulation*, as its dual is a path.[1] Unfortunately, a Hamiltonian triangulation does not always exist, see Fig. 7. Either of the two robot enhancements which we have introduced allows robots to use additional points of the boundary of P to compute a baseline triangulation.

Fig. 6. At v, a robot chooses vw as the direction of the robot's walk. After reaching w, it can continue in the same direction until it hits the boundary at point w'.

Fig. 7. A polygon and its unique triangulation with triangles specified solely of vertices of the polygon. The triangulation is depicted by dashed lines.

In the case of a baseline triangulation a robot can count the targets with the following algorithm. For every baseline triangle the robot moves to the vertex of the triangle (recall that this might not be a vertex of the polygon) opposite to the baseline edge, and counts the targets that are visible between the two vertices of the edge. Clearly, in this way every target is counted exactly once. Hence, a general algorithm that allows a robot to solve the counting problem is as follows: it is composed of a procedure to produce a baseline triangulation and of a navigation scheme to visit every baseline triangle exactly once.

Narasimhan [10] presents an algorithm that recognizes whether a polygon has a Hamiltonian triangulation and computes one. The algorithm can be adapted for a robot that can discern convex vertices from reflex vertices [9] (which is not directly possible in our model [1]). Hence, such a robot can resolve whether a polygon admits a Hamiltonian triangulation and exactly count the number of targets in that case. However, when the polygon is non-Hamiltonian, this approach does not give anything useful, whereas our scheme, given in the following section, works for general polygons.

3.3 Walking Along Edge- and Diagonal-Extensions

In this section we consider robots that can walk along edge- and diagonal-extensions. We show that such robots can partition any simple polygon into baseline triangles, and thus can count the targets.

---

[1] A dual of a triangulation is a graph, where each triangle corresponds to one vertex and there is an edge between two vertices iff the two corresponding triangles share an edge in the triangulation.

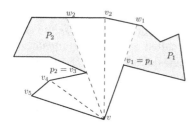

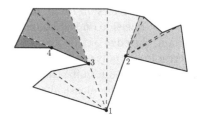

Fig. 8. The extensions of a vertex v define baseline triangles and pockets of v

Fig. 9. The resulting baseline triangulation of the algorithm and the visited pockets (grey). The labeled dots show the order of the recursion calles.

Consider a robot at a vertex v of the polygon P. Let $v_1, v_2, \ldots, v_i \ldots$ denote the visible vertices from v, cyclically ordered in the counterclockwise direction. Observe that the lines vv_i partition the visible part of P into baseline triangles (all with a common point v), each with at least one baseline edge. See Fig. 8 for an illustration, where the triangles vw_1v_2, vv_2w_2, vv_3v_4 and vv_4v_5 partition the visible part of P. Thus we can partition the visible part of P. Observe that the invisible part of P is a set of disjoint simple sub-polygons. In the example from Fig. 8 the sub-polygons P_1 and P_2 form the invisible part of P. We call such a sub-polygon a *pocket* of P. Observe that a pocket is created by a line which is an edge-extension or a diagonal-extension. Applying a recursive partitioning approach on the pockets, we create a partition of P into triangles with at least one edge on the boundary of P (see Fig. 9). Let T denote this triangulation.

The main idea of the algorithm is to count all targets from the robot's position v and then proceed recursively in the corresponding pockets of the polygon, thus navigating through T and counting the targets in the triangles of T. We begin with a high-level description. For a vertex v let $P_1, \ldots, P_\ell$ denote pockets of P defined by all extensions originating at v. Let p_i, $i = 1, \ldots, \ell$, denote the visible vertex whose extension defines P_i. Let w_i be the point of P for which vw_i is the extension of vp_i.

Counting in Simple Polygons

1. Count all the targets that are visible from the robot's position at vertex v.
2. Put a pebble at v and remember the position of v in the respective piv of every vertex p_i and of every point w_i.
3. Recursively count the targets in P_i, $i = 1, \ldots, \ell$, by marking the point w_i with a pebble and going to p_i.

When a robot walks to vertex p_i to start a recursive call for pocket P_i, it first checks the position of the pebble that marks the point w_i. Next the robot determines which vertices (and targets) visible from p_i belong to pocket P_i. Let k be the number of vertices (including w_i) and targets visible from p_i. Let h be the index of w_i in the piv of vertex p_i. If pocket P_i lies to the right of p_iw_i, then

P_i contains the vertices and targets from the piv of p_i with index $1, 2, \ldots, h$. If pocket P_i lies to the left of $p_i w_i$, then P_i contains the vertices and targets from the piv of p_i with index $h, h+1, \ldots, k$. Observe that P_i lies to the right of $p_i w_i$ if and only if p_i is the first end-point of the diagonal in piv of vertex v.

The robot at vertex p_i knows which part of its piv represents the sub-polygon P_i and it can therefore perform the same steps of the **Counting in Simple Polygons** algorithm on the pocket P_i only. Before that, the pebble from w_i is collected as it is no longer needed. When the robot finishes the counting in P_i it returns to the vertex v (using the stored navigation information) and continues there.

Theorem 4. *A robot with one pebble, able to walk along extensions, can count the number of targets in a simple polygon of n vertices in $O(n)$ steps with $O(n)$ memory.*

Proof. Let T be the baseline triangulation of P produced by the algorithm. The triangles of T are defined by vertices of P and intersection points between polygonal edges and extensions. Observe first that the algorithm provides a consistent navigation scheme through T. Thus every target is counted exactly once. Note that the dual of T is a tree. The edges of every triangle of T contain three vertices of P, which are then mutually visible and build a triangle. Since any triangulation of a polygon (in the classical sense) has exactly $n-2$ triangles, T can only be smaller and the dual of T has $O(n)$ vertices, which is also the number of steps of the algorithm (since the robot spends a constant number of steps in every triangle of T). The robot stores the necessary information to return from a recursive call – the predecessor v of every vertex p_i. Hence, $O(n)$ memory is sufficient. □

3.4   Walking with a Compass

In this section we consider in addition one fixed direction in which the robot can move. Without loss of generality, we assume that a robot sitting at a vertex can, additionally to moving to all visible vertices, move also along the vertical line going through the robot's position.

We present an algorithm that computes a baseline triangulation in any simply- or multiply-connected polygon and navigates the robot such that each triangle is considered for counting exactly once and thus it allows the robot to count the number of targets in the polygon. To simplify the presentation we first use an arbitrary number of pebbles – we show later how to use only a constant number of pebbles.

The key observation is that all the vertical extensions from vertices of a polygon P partition the polygon into baseline triangles and convex quadrilaterals for which two opposite sides are on the boundary of P (Fig. 10). Each quadrilateral can be subsequently partitioned into two baseline triangles (by picking a diagonal as the common boundary of the triangles).

Hence, using at most $2n$ pebbles, the robot can mark every end-point of every vertical extension which then imposes a baseline triangulation. This can be done

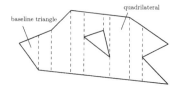

Fig. 10. A multi-connected polygon with its partition by vertical extensions

by visiting every vertex of P (using one pebble) [1]. To count every target exactly once, the robot goes through every vertex or pebble p and considers only triangles lying *above* p and on its *right* (if any). Since every triangle has one vertical side, the robot can always reach the opposite vertex of the baseline side in one step and count the targets in the triangle, and return back.

We now show how to reduce the number of used pebbles at the cost of an increased number of steps. The robot does not mark all the quadrilaterals at once, but one by one. Let us call an endpoint of a vertical extension a *q-node*. We show how to navigate through all the vertices and q-nodes in a consistent way. We begin with the navigation through vertices of P from [1] (the navigation can be computed in $O(n^3)$ steps with $O(n)$ memory), where every edge of P is visited exactly once. If a robot moves in this navigation along a polygonal edge uv, we compute all the q-nodes lying on this edge and before the robot moves to v it visits all the q-nodes in the order of increasing distance from u.

Let us consider the situation where the robot is at a point p (a vertex u or a q-node) of the edge uv and it wants to move to the next q-node. The robot can find the next q-node by sequentially creating all q-nodes (by going to every vertex of P) and checking which one lies on the edge uv and closest to p. Specifically, using a pebble the robot marks the initial position p. The next pebble is used to mark the so-far closest q-node on the edge uv. The robot goes through every vertex w of P and creates q-nodes lying on the vertical extensions of w. For every such q-node the robot checks whether it lies on the edge uv and whether it is closer to u than the current best. The two pebbles make this operation easy for the robot.

Theorem 5. *A robot with 2 pebbles, able to walk along vertical extensions, can count the number of targets in a polygon with n vertices in $O(n^3)$ steps and with $O(n)$ memory.*

4   Conclusions

We considered a minimalistic computational framework of mobile sensors – simple robots, whose visibility-based sensing reflects just the combinatorial character of the environment. We investigated their capabilities on the problem of counting points of interest (targets) in a polygon P and considered two scenarios. We have shown that in the friendly environment the robots can count the targets using two pebbles. In the hostile environment the robots cannot count the targets and they cannot even approximate the number of targets by a multiplicative factor less than 2. We have looked at possible minimum extensions of the robots' capabilities that allow to count targets. We have considered two such extensions – walking along edge- and diagonal-extensions, and walking with compass.

We have not answered all interesting questions and many of these remain open for the future research. For example, what is the best approximation ratio of the problem? Is the lower bound tight or is there a better approximation algorithm? What is the inherent power of pebbles: can we do anything without them? Are there simpler robots' enhancements that allow the robots to count the targets? Can a collaboration of more robots do better than a single robot?

References

1. Suri, S., Vicari, E., Widmayer, P.: Simple robots with minimal sensing: From local visibility to global geometry. In: Proceedings of the Twenty-Second National Conference on Artificial Intelligence and the Nineteenth Innovative Applications of Artificial Intelligence Conference, pp. 1114–1120. AAAI Press, Stanford, California, USA (2007)
2. Chvátal, V.: A combinatorial theorem in plane geometry. Journal of Combinatorial Theory 18, 39–41 (1975)
3. Latombe, J.C.: Robot Motion Planning. Kluwer Academic Publishers, Norwell, MA, USA (1991)
4. LaValle, S.M.: Planning Algorithms. Cambridge University Press, Cambridge, UK (2006)
5. Tovar, B., Freda, L., LaValle, S.M.: Using a robot to learn geometric information from permutations of landmarks. Contemporary Mathematics (to appear, 2007)
6. Yershova, A., Tovar, B., Ghrist, R., LaValle, S.M.: Bitbots: Simple robots solving complex tasks. In: Proceedings of the Twentieth National Conference on Artificial Intelligence and the Seventeenth Innovative Applications of Artificial Intelligence Conference, pp. 1336–1341 (2005)
7. Guibas, L.J., Latombe, J.C., LaValle, S.M., Lin, D., Motwani, R.: A visibility-based pursuit-evasion problem. International Journal of Computational Geometry and Applications 9, 471–494 (1999)
8. Sachs, S., Rajko, S., LaValle, S.M.: Visibility-based pursuit-evasion in an unknown planar environment. International Journal of Robotics Research 23, 3–26 (2004)
9. Gfeller, B., Suri, S., Mihalák, M., Vicari, E., Widmayer, P.: Counting targets with mobile sensors in an unknown environment. Technical Report 571, Department of Computer Science, ETH Zurich (2007)
10. Narasimhan, G.: On hamiltonian triangulations in simple polygons. International Journal of Computational Geometry and Applications 9, 261–275 (1999)

Asynchronous Training in Wireless Sensor Networks

Ferruccio Barsi[1], Alan A. Bertossi[2], Francesco Betti Sorbelli[1], Roberto Ciotti[1], Stephan Olariu[3], and Cristina M. Pinotti[1]

[1] Department of Computer Science and Mathematics, University of Perugia, 06123 Perugia, Italy
{barsi,pinotti}@unipg.it
[2] Department of Computer Science, University of Bologna, Mura Anteo Zamboni 7, 40127 Bologna, Italy
bertossi@cs.unibo.it
[3] Department of Computer Science, Old Dominion University, Norfolk, VA 23529-0162, USA
olariu@cs.odu.edu

Abstract. A scalable energy-efficient training protocol is proposed for massively-deployed sensor networks, where sensors are initially anonymous and unaware of their location. The protocol is based on an intuitive coordinate system imposed onto the deployment area which partitions the anonymous sensors into clusters. The protocol is asynchronous, in the sense that the sensors wake up for the first time at random, then alternate between sleep and awake periods both of fixed length, and no explicit synchronization is performed between them and the sink. Theoretical properties are stated under which the training of all the sensors is possible. Moreover, a worst-case analysis as well as an experimental evaluation of the performance is presented, showing that the protocol is lightweight and flexible.

1 Introduction

Ultra-high integration and low-power electronics have enabled the development of miniaturized, low-cost, battery-operated sensor nodes (*sensors*, for short) that integrate signal processing and wireless communications capabilities [1,14]. Many applications require the aggregation of massively deployed sensors into sophisticated infrastructures, called *sensor networks*. Recently, it has been recognized that it would be beneficial to augment the sensor networks by more powerful entities, called *sinks*. While the sensors are tasked mainly to sense their immediate neighbourhood, the sinks collect, aggregate and fuse the data harvested by the sensors in order to act on the environment in a meaningful way. The typical mode of operation of a sink is to task the sensors in a portion of a disk of radius ρ centered at itself to produce data relevant to the mission at hand. Once this data has been aggregated, the sink has a good idea of what action to take. For instance, Figure 1(a) and Figure 1(b) illustrate, respectively, a disk around a sink and a disk subdivided in portions.

M. Kutyłowski et al. (Eds.): ALGOSENSORS 2007, LNCS 4837, pp. 46–57, 2008.
© Springer-Verlag Berlin Heidelberg 2008

The massive deployment of tiny sensors results in sensors initially unaware of their location. However, many probable applications as well as the sink assignment of tasks to the sensors require that individual sensors have to determine either their exact geographic location or else a coarse-grain approximation thereof. The former task is referred to as *localization* and has been extensively studied in the literature [8,10]. The latter task, referred to as *training*, has been considered in several recent papers by Olariu *et al.* [3,11,12,13]. In particular, they devised some training protocols for sensor networks, which differ on whether or not sensors and sink need some kind of explicit synchronization. Such training protocols have different performance, measured in terms of total time for training, overall sensor awake time, and number of sensor sleep/wake transitions.

The main contribution of this paper is to further study the task of training, assuming the same asynchronous model as that originally defined in [13]. In particular, the model in [13] assumes that the sink and the sensors are asynchronous, in the sense that the sensors wake up for the first time at random and then alternate between sleep and awake periods both of fixed length, while no explicit synchronization is performed between them and the sink. The present paper completes the work of [13], by stating novel theoretical properties of the parameters of the training protocol under which the training of all the sensors in the network is possible. Moreover, improvements of the protocol are presented which are lightweight in terms of both the number of wake/sleep transitions and the overall sensor awake time for training.

The remainder of this paper is organized as follows. Section 2 discusses the wireless sensor network model and introduces the task of training. Training imposes a coordinate system which divides the sensor network area into equiangular wedges and concentric coronas centered at the sink, as first suggested in [12]. Section 3 is the backbone of the entire paper, presenting the theoretical underpinnings of the training protocol, along with its worst-case performance analysis. Section 4 presents an experimental evaluation of the performance, tested on randomly generated instances, showing that the protocol behaves much better in the average case than in the worst case. Finally, Section 5 offers concluding remarks.

2   The Network Model

In this work a wireless sensor network is assumed that consists of a sink and a set of sensors randomly deployed in its broadcast range as illustrated in Figure 1(a). For simplicity, the sink is centrally placed, although this is not really necessary.

A sensor is a device that possesses three basic capabilities: sensory, computation, and wireless communication, and operates subject to the following fundamental constraints:

a. Each sensor has a modest non-renewable energy budget and a transmission range of r;
b. In order to save energy, each sensor alternates between *sleep* periods and *awake* periods – the sensor sleep-awake cycle is of total length L out of which the sensor is in sleep mode for $L - d$ time and in awake mode for d time;

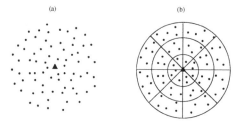

(a)   (b)

Fig. 1. (a)The sensors in a disk centered at a sink. (b) The disk subdivided in portions.

c. Each sensor is asynchronous – it wakes up for the first time according to its internal clock and is not engaging in an explicit synchronization protocol with either the sink or the other sensors;
d. Each sensor has no global information about the network topology, but can hear transmissions from the sink;
e. Sensors are *anonymous* – to assume the simplest sensor model, sensors do not need individually unique IDs;
f. Individual sensors must work *unattended* – once deployed it is either infeasible or impractical to devote attention to individual sensors.

The task of training is essential in several applications. One example is clustering where the set of sensors deployed in an area is partitioned into clusters [1,2,5]. As a result of training, we impose a coordinate system onto the sensor network in such a way that each sensor belongs to exactly one cluster. The coordinate system involves establishing [12]:

1. *Coronas*: The deployment area is covered by k coronas $c_0, c_1, \ldots, c_{k-1}$ determined by k concentric circles, centered at the sink, whose radii are $0 < r_0 < r_1 < \cdots < r_{k-1} = \rho$;
2. *Wedges*: The deployment area is ruled into a number of equiangular wedges, centered at the sink, which are established by directional transmission [11].

For the sake of simplicity, in this paper, it is assumed that the corona width is equal to the sensor transmission range r, and hence the (outer) radius r_i of corona c_i is equal to $(i + 1)r$. As illustrated in Figure 1(b), at the end of the training period each sensor has acquired two coordinates: the identity of the corona in which it lies, as well as the identity of the wedge to which it belongs. In particular, a cluster is the locus of all nodes having the same coordinates in the coordinate systems [11].

3   The Corona Training Protocol

The main goal of this section is to present the details of the corona training protocol (the wedge training protocol is similar and will not be discussed), where each individual sensor has to learn the identity of the corona to which it belongs, regardless of the moment when it wakes up for the first time. To see how this is done, it is useful to assume the time ruled into slots. The sensors and the sink use equally long, in phase slots, but they do not necessarily start counting the time from the same slot.

The idea of the protocol, called Flat–, is as follows. Immediately after deployment the sink cyclically repeats a transmission cycle which involves k broadcasts at successively lower power levels. Each broadcast lasts for a slot and transmits a beacon equal to the identity of the outmost corona reached. Precisely, the sink starts out by transmitting the beacon $k - 1$ at the highest power, sufficient to reach the sensors up to the outmost corona c_{k-1}; next, the sink transmits the beacon $k - 2$ at a power level that can be received up to corona c_{k-2}, but not by the sensors in corona c_{k-1}. For the subsequent $k - 2$ slots, the sink continues to transmit at decreasing power levels until it concludes its transmission cycle with a broadcast that can be received only by the sensors in corona c_0. In general, at time slot τ, with $\tau \geq 0$, the sink transmits the beacon $k - 1 - |\tau|_k$ with a power level that can reach all the sensors up to corona $c_{k-1-|\tau|_k}$, where $|a|_b$ stands for the non negative remainder of the integer division between a and b (i.e. $|a|_b$ is the same as a modulo b). The sink transmission cycle is repeated for a time sufficient to accomplish the entire corona training protocol.

In order to describe the Flat– protocol for sensors, it is crucial to point out that each sensor is aware of the sink behaviour and of the total number k of coronas. Immediately after deployment, each sensor wakes up at random within the 0-th and the $(k - 1)$-th time slot and starts listening to the sink for d time slots (that is, its awake period). Then, the sensor goes back to sleep for $L - d$ time slots (that is, its sleep period). Such a sleep/wake transition will be repeated until the sensor will learn the identity of the corona to which belongs, that is, until the sensor will be trained. Each sensor, during the training process, uses a k-bit register R to keep track of the beacons, i.e. corona identities, transmitted by the sink while the sensor is awake. As soon as the sensor hears a sink transmission for the first time, it starts to fill the register R and it is able to learn the sink global time t within the current sink transmission cycle, that is $t = |\tau|_k$. From now on, such a time will regularly increase so that the sensor can derive from t the beacon $|k - 1 - t|_k$ that the sink is transmitting. Then, in each time slot when the sensor is awake, one entry of R can be always set either to 0 or to 1. In fact, if the sensor hears beacon c, then it sets $R_c = 1$, while if the sensor hears nothing, it sets $R_{|k-1-t|_k} = 0$. Note that the awake sensors which belong to corona c, with $c > 0$, are able to receive any transmitted beacon from c up to $k-1$, whereas they cannot hear the beacons from 0 up to $c-1$. Hence, if a sensor sets $R_c = 0$ (resp., $R_c = 1$) then it belongs to a corona whose identity is higher than (resp., smaller than or equal to) c. Note that only the sensors in corona 0 can hear beacon 0 and thus they are the only ones which can set $R_0 = 1$. From the above discussion, the following *training condition* holds:

Lemma 1. *A sensor which belongs to corona c, with $c > 0$, is trained as soon as the entries R_c and R_{c-1} of its register R are set to 1 and 0, respectively. A sensor which is in corona 0 is trained as soon as R_0 is set to 1.*

In the following, some conditions on the parameters k, L, and d will be investigated which guarantee that all the sensors are trained, independent of their first wakeup time and from the corona c they belong to. Hereafter, let (a, b) denote the *greatest common divisor* between a and b. Moreover, if $(a, b) = 1$, let $\left|\frac{1}{a}\right|_b$ be the multiplicative inverse of a modulo b (e.g. see [7]).

Consider a sensor that wakes up for the first time at the global time slot $\tau = x$, while the sink is transmitting the beacon $c_x = |k - 1 - \tau|_k = |k - 1 - x|_k$. The i-th sleep-awake cycle of such a sensor starts at time $x + iL$ while the sink is transmitting the beacon $|k - 1 - x - iL|_k = |c_x - i|L|_k|_k$, with $i \geq 0$. Observe that L and k can be rewritten as $L = gL'$ and $k = gk'$, where $g = (L, k)$. Since $|L|_k = |gL'|_k = g|L'|_{k'}$, one has $|c_x - i|L|_k|_k = |c_x - ig|L'|_{k'}|_k$. Hence, there are exactly k' different coronas which can be transmitted by the sink when the sensor starts its awake period, independent of how long the training process will be. Indeed, since $|c_x - (i + k')|L|_k|_k = |c_x - (i + k')g|L'|_{k'}|_k = |c_x - (ig + k'g)|L'|_{k'}|_k = |c_x - (ig + k)|L'|_{k'}|_k = |c_x - ig|L'|_{k'}|_k$, the same corona is transmitted again at the beginning of any two awake periods of the sensor which are k' apart. Moreover, for any two awake periods, say the i-th and the j-th ones, such that $i > j$ and $i - j < k'$, the coronas c_{x+iL} and c_{x+jL} are distinct and differ by a multiple of g. Such overall k' corona identities can be rearranged so that in the new order two consecutive coronas differ exactly by g. Indeed the s-th corona in the new order, that is $|c_x - sg|_k$, corresponds to the first beacon transmitted in the $\left|s\left|\frac{1}{L'}\right|_{k'}\right|_{k'}$-th awake period, with $0 \leq s \leq k' - 1$. Therefore, after exactly k' sleep-awake cycles, and hence $k'L$ time slots, the sink has performed $\frac{k'L}{k} = \frac{k'L}{gk'} = \frac{L}{g} = L'$ transmission cycles. From now on, the behaviour of the sensor and the sink is cyclically repeated with a period of $k'L$ time slots. In other words, in the k'-th awake period, the sensor and the sink are in the same reciprocal state as in the 0-th one, the only difference being that the sensor has heard the sink at least once. Summarizing:

Lemma 2. *Fixed L, d, and k, there are exactly $k' = \frac{k}{(L,k)}$ different corona identities that can be transmitted by the sink when the sensor starts any awake period. Assuming that the sensor wakes up for the first time at slot x, $0 \leq x \leq k - 1$, then the corona identity transmitted when the sensor starts its i-th awake period is $|c_x - i(L, k)|L'|_{k'}|_k = |c_x - |i|_{k'}(L, k)|L'|_{k'}|_k$. Such k' coronas identities can be reindexed as $|c_x - s(L, k)|_k$, for $0 \leq s \leq k' - 1$.*

Thus, since during an awake period of the sensor the sink transmits d distinct beacons, overall the sink transmits no more than $k'd$ different corona identities during the first k' awake periods of the sensor, and such coronas will be repeatedly transmitted again. Recalling that a sensor starts to fill R only after it heard the sink for the first time and observing that all the entries that the sensor can fill are set in further k' sleep-awake cycles, it follows:

Lemma 3. *Fixed L, d, and k, all the entries of R the sensor can fill are set within the first $\frac{2k}{(L,k)}$ sleep-awake cycles, or equivalently, $\frac{2L}{(L,k)}$ sink transmission cycles.*

In other words, if a sensor has not been trained after $\frac{2kL}{(L,k)}$ time slots, it will never be trained, independent of how long the training process will continue.

Theorem 1. *The training condition is satisfied for all the sensors after at most* $2k' = 2\frac{k}{(L,k)}$ *sleep/wake cycles if and only if* $d \geq (L,k)$.

Proof. For brevity let $g = (L,k)$. By contradiction, suppose that all the sensors have been trained and let $d < g$. By Lemmas 2 and 3, after at most $\frac{2k}{g}$ sleep-awake periods, each sensor has filled at most $k'd$ entries of R. Since $d < g$, each sensor has filled less than k entries of R. Such filled entries depend on the time slot x when the sensor woke up for the first time. Consider now all the sensors that woke up at the same time x. Note that they have filled, although with different configurations, the same positions of R independent of the corona they belong. Let c be one unfilled entry of R. By the hypothesis of massive random deployment, there is at least one sensor that woke up at time x in each corona, and hence at least one sensor in corona c. Clearly, such a sensor will not be trained because the training condition in Lemma 1 will be never satisfied.

Conversely, if $d \geq g$, by Lemma 2, in k' consecutive sleep-awake cycles, the beacons transmitted by the sink in the first slot of such k' cycles are exactly g apart. Since an awake period lasts $d \geq g$ slots, at least g new corona identities are transmitted by the sink during an awake period of the sensor. Hence, after the first k' awake periods, the sensor fills at least g entries of R in each awake period and completely fills R in at most other k' awake periods. Therefore, the sensor is trained in at most $2k'$ consecutive awake periods by Lemma 3. Note that this happens for all the sensors, independent of their first wake-up time x and of the corona c to which they belong.

In the following, some properties of the training protocol are analyzed starting from a couple of particular cases, namely, when $d = (L,k)$ and $d = |L|_k$. Note that, since $d = |L|_k = (L,k)|L'|_{k'}$, Theorem 1 holds in both cases.

First, the maximum number of sleep/wake transitions required to train a sensor is discussed. Precisely, the following lemma specifies when a sensor, that wakes up for the first time at slot x, is awake while the sink is transmitting c.

Lemma 4. *Let c be any corona identity and assume $d = (L,k)$. The sink transmits the beacon c during the $i_{c,x}$-th awake period of a sensor that wakes up for the first time at slot x, where $i_{c,x} = \left|\left|\left\lfloor \frac{|c_x - c|_k}{d} \right\rfloor \right| \frac{1}{L'}|_{k'}\right|_{k'}$, $L' = \frac{L}{d}$, and $k' = \frac{k}{d}$.*

Proof. When the sensor wakes up at time x the sink is transmitting the beacon c_x. Moreover, the beacon values decrease within a sink transmission cycle. Thus, the beacon c will be transmitted, starting from c_x, during the j-th group of d consecutive corona identities such that $j = \left\lfloor \frac{|c_x - c|_k}{d} \right\rfloor$. Such a j-th group of d consecutive corona identities will be transmitted during the $i_{c,x}$-th sensor awake period in which the sink transmits $\left|c_x - \left\lfloor \frac{|c_x - c|_k}{d} \right\rfloor d\right|_k$ as the first beacon. Hence, by Lemma 2, $i_{c,x}$ is derived by solving the equation $|c_x - i_{c,x}(L,k)|L'|_{k'}|_k$ $= \left|c_x - \left\lfloor \frac{|c_x - c|_k}{d} \right\rfloor d\right|_k$. Recalling that $d = (L,k)$, the solution of the equation is $i_{c,x} = \left|\left|\left\lfloor \frac{|c_x - c|_k}{d} \right\rfloor \right| \frac{1}{L'}|_{k'}\right|_{k'}$.

Lemma 5. *Let c be any corona identity and assume $d = |L|_k$. The sink transmits the beacon c during the $i_{c,x}$-th awake period of a sensor which wakes up for the first time at slot x, where $i_{c,x} = \left\lfloor \frac{|c_x - c|_k}{d} \right\rfloor$.*

Theorem 2. *Let $(L, k) \leq d < |L|_k$. A sensor which wakes up for the first time at slot x and belongs to corona c, with $c > 0$, is trained during the i-th awake period where $i = i_{c-1,x}$, if $i_{c,x} \leq i_{c-1,x}$, or $i \leq i_{c,x} + \left|\frac{1}{L'}\right|_{k'}$, if $i_{c,x} > i_{c-1,x}$. If $c = 0$, then $i = i_{0,x}$.*

Proof. Consider first the case $d = (L, k)$. If $i_{c,x} \leq i_{c-1,x}$, during the $i_{c,x}$ awake period the sensor hears the beacon c and hence it sets $R_c = 1$. Moreover, during the $i_{c-1,x}$ awake period, the sensor sets $R_{c-1} = 0$ because it does not hear $c - 1$ but, having already heard c, it knows what the sink is transmitting. If $i_{c,x} > i_{c-1,x}$, in the worst case the sensor hears for the first time during the $i_{c,x}$-th awake period and sets $R_c = 1$. Then, the beacon $c - 1$ will be transmitted at the i-th awake period such that $|c_x - i(L, k)|L'|_{k'}|_k = |c_x - (j+1)d|_k$, where $j = \left\lfloor \frac{|c_x - c|_k}{d} \right\rfloor$. Solving the above equation, one has $i = \left|(j+1)\left|\frac{1}{L'}\right|_{k'}\right|_{k'} = i_{c,x} + \left|\frac{1}{L'}\right|_{k'}$. When $d > (L, k)$, since by Lemma 2 the k' coronas transmitted by the sink when the sensor wakes up do not depend on d, the sensor cannot be trained later than in the case $d = (L, k)$. $\square$

Theorem 3. *Let $|L|_k \leq d < k$. A sensor which wakes up for the first time at slot x and belongs to corona c, with $c > 0$, is trained during the i-th awake period where $i = i_{c-1,x}$, if $i_{c,x} \leq i_{c-1,x}$, or $i \leq i_{c,x} + 1$, if $i_{c,x} > i_{c-1,x}$. If $c = 0$, then $i = i_{0,x}$.*

In order to analytically evaluate the performance of the Flat– training protocol, let us consider the number ν of sensor sleep/wake transitions, the overall sensor awake time ω, and the total time τ for training. Since a sleep-awake period has length L, and a sensor is awake for d time slots per sleep-awake period, one has $\omega = \nu d$ and $\tau = \nu L$. Thus, the worst case performance for the Flat– protocol can be summarized as follows:

Corollary 1. *Fixed L, d, and k, if $d < (L, k)$ then there are sensors which cannot be trained by the Flat– protocol; otherwise all the sensors are trained, and:*

1. *If $(L, k) \leq d < |L|_k$, then $\nu \leq \frac{k}{(L,k)} + \left|\frac{1}{L'}\right|_{k'}$, where $k' = \frac{k}{(L,k)}$ and $L' = \frac{L}{(L,k)}$;*
2. *If $|L|_k \leq d < k$, then $\nu \leq \left\lfloor \frac{k}{|L|_k} \right\rfloor + 1$,*
3. *If $d = k$, then $\nu \leq 2$.*

3.1   Improvements

The Flat– protocol can be improved so as to reduce the number ν of sleep/wake transitions, and hence also the overall sensor awake time as well as the total time for training.

In fact, as soon as a sensor hears the sink transmission for the first time, it learns from the beacon the sink global time modulo the sink transmission cycle. Therefore, it can immediately retrieve backwards the coronas which it did not hear and which were transmitted by the sink during its previous awake periods, setting to 0 the corresponding entries of R. The resulting improved protocol is called $Flat$. Observed that the sensor behaviour is the same as it would have set the entries of R since its first wake up, Lemma 3 and Theorem 1 can be restated as follows:

Lemma 6. *Fixed L, d, and k, all the entries of R the sensor can fill are set within the first $\frac{k}{(L,k)}$ sleep-awake cycles, or equivalently, $\frac{L}{(L,k)}$ sink transmission cycles.*

Theorem 4. *The training condition is satisfied for all the sensors after at most $k' = \frac{k}{(L,k)}$ sleep/wake cycles if and only if $d \geq (L,k)$.*

In other words, after at most $k'L$ time slots the training process is completed. Such a bound is tight in the particular case that $d = (L,k)$, while it can be lowered when $d = |L|_k$. Indeed, Theorems 2 and 3 become:

Theorem 5. *A sensor which wakes up for the first time at slot x and belongs to corona c is trained during the i-th awake period where $i = max\{i_{c-1,x}, i_{c,x}\}$, if $c > 0$, or $i = i_{0,x}$, if $c = 0$.*

Note that i varies between 0 and $\frac{k}{|L|_k}$ when $d \geq |L|_k$, whereas it varies between 0 and $\frac{k}{(L,k)}$ otherwise. Hence, the worst case performance for the Flat protocol is summarized below:

Corollary 2. *Fixed L, d, and k, if $d < (L,k)$ then there are sensors which cannot be trained by the Flat protocol; otherwise all the sensors are trained, and:*

1. *If $(L,k) \leq d < |L|_k$, then $\nu \leq \frac{k}{(L,k)}$;*
2. *If $|L|_k \leq d < k$, then $\nu \leq \lceil \frac{k}{|L|_k} \rceil$;*
3. *If $d = k$, then $\nu = 1$.*

Note that, when $d = (L,k)$ or $d = |L|_k$, each of the k distinct beacons is transmitted exactly once in the $\lceil \frac{k}{d} \rceil$ awake periods during which each sensor is trained.

A further improvement to the Flat protocol exploits the fact that when a sensor hears a beacon c, it knows that it will also hear all the beacons greater than c, and thus it can immediately set to 1 the entries from R_c up to R_{k-1}. Similarly, when a sensor sets an entry R_c to 0, it knows that it cannot hear any beacon smaller than c, and thus it can immediately set to 0 the entries from R_{c-1} down to R_0, too. In contrast to the previous protocols, the sensor now fills entries of R relative to beacons not yet transmitted during its awake periods. Therefore, it can look ahead to decide whether it is worthy or not to wake up in the next awake period. If the d entries of R that will be transmitted by the sink in the next awake period have already been filled, then the sensor can skip its next awake period, thus saving energy. The sensor repeats the look ahead process above until at least one unfilled entry is detected among the d entries corresponding to a future awake period. The resulting protocol is called *Flat+*. Clearly, the number ν of sleep/wake transitions of Flat+ cannot be larger than that of Flat. Moreover, when $d = |L|_k$ or $d = (L, k)$, one can find bad instances where ν, in the worst case, is the same for both Flat+ and Flat. For example, when $d = |L|_k$, a sensor which belongs to corona c and wakes up when the sink transmits $c_x = c - 1$ requires $\lceil \frac{k}{d} \rceil$ transitions to be trained by both protocols. However, as it will be experimentally checked in the following section, the average behaviour of Flat+ is much better than that of Flat.

4   Experimental Tests

In this section, the worst and average performance of the Flat–, Flat, and Flat+ protocols are experimentally tested. In the simulation, the number k of coronas is fixed to 64, and each corona has a unit width. There are $N = 10000$ sensors uniformly distributed within a circle, centered at the sink, having radius $\rho = k$. Precisely, the polar coordinates of each sensor are generated choosing at random two real numbers. The first one, uniformly distributed between 0 and k, represents the radial coordinate of the sensor, that is, its distance from the sink. The second number, uniformly distributed between 0 and 2π, represents the angular coordinate of the sensor, that is, the positive angle required to reach the sensor from the polar axis. The length L of the sensor sleep-awake cycle assumes the values 104 and 168. Finally, in all the experiments, the sensor awake period d is an integer that varies, with a step of 4, between the greatest common divisor $(L, k) = 8$ and $k = 64$, thus including $|L|_k = 40$. The results are reported only when all the sensors can be trained, that is for $d \geq 8$, and are averaged over 3 independent experiments. In the experiments, both the worst and average number of transitions, denoted by $\nu_{\max}$ and ν_{avg}, as well as both the worst and average overall sensor awake time, $\omega_{\max}$ and ω_{avg}, are evaluated. Such average values are obtained by summing up the values for each single sensor and then dividing by the number of sensors. Moreover, the total time τ, which measures the time required to terminate the whole training process, is evaluated.

Figure 2 shows the number $\nu_{\max}$ and ν_{avg} of transitions for the different values of d. According to Corollaries 1 and 2, Flat– has $\nu_{\max} = 13$ when $d = 8$, while both Flat and Flat+ have $\nu_{\max} = 8$. Similarly, when $d = 40$, all protocols take $\nu_{\max} = 2$ transitions. Except for the extreme values $d = 8$ and $d = 64$, the greatest percentage of gain for $\nu_{\max}$ is achieved when $d = 24$, where both Flat+ and Flat employ forty percent less transitions than Flat–. As regard to the average performance, one notes that ν_{avg} is considerable better than $\nu_{\max}$ for all three protocols. Flat and Flat– have almost the same average performances, while Flat+ always behaves better than them. In particular, its greatest percentage of gain for ν_{avg} is obtained in the range $8 \leq d \leq 20$, where Flat+ improves about twenty/thirty percent upon Flat–.

Figure 3 shows the awake times $\omega_{\max} = \nu_{\max} d$ and $\omega_{\mathrm{avg}} = \nu_{\mathrm{avg}} d$, which measure the overall energy spent by each sensor to be trained. Although the

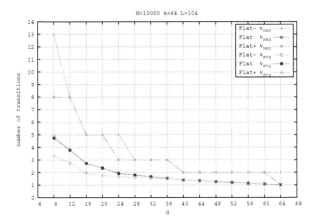

Fig. 2. Number of transitions when $k = 64$, $L = 104$, and $8 \leq d \leq 64$

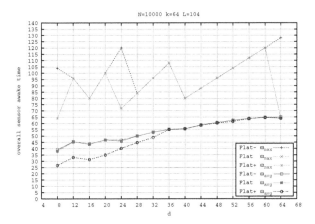

Fig. 3. Overall sensor awake time when $k = 64$, $L = 104$, and $8 \leq d \leq 64$

number of transitions decreases as d increases, Figure 3 suggests to choose a small value of d from the sensor awake time perspective. The minimum ω_{max} is achieved by Flat and Flat+ for $d = 8$ and $d = 64$, as expected by Corollaries 1 and 2. However, when $d = 8$, ω_{avg} lowers to about two thirds of ω_{max} for Flat– and Flat, and to about one third for Flat+. Note that Flat+ has the maximum gain when d is small. Indeed, it can fill the same entries of R just listening to the sink for a single slot or for d slots. Hence, small values of d save the same number of transitions as larger values, but allow sensors to reduce their energy consumption because they stay awake for smaller periods.

Figure 4 exhibits the total time τ required to accomplish the entire training task, for both $L = 104$ and $L = 168$. Since $|168|_{64} = |104|_{64} = 40$, by Lemma 2, each protocol maintains the same behaviour with respect to the number of

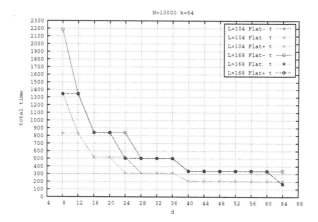

Fig. 4. Total time for training when $k = 64$, $L = 104$ or $L = 168$, and $8 \leq d \leq 64$

transitions. Thus, the plots for $L = 168$ of $\nu_{\max}$ and ν_{avg}, and hence of $\omega_{\max}$ and ω_{avg}, are exactly the same as those shown in Figures 2 and 3. Recalling that $\tau = \nu_{max}L$, the total time for $L = 168$ scales by a constant $\frac{168}{104}$, as depicted in Figure 4. In general, all values of L such that $|L|_k$ is the same present the properties above, namely, ν and ω are identical, while τ scales. Therefore, the minimum total time τ is achieved for the smallest value of L. However, larger values of L could be also selected in order to increase the longevity of the wireless sensor network. Fixed d, a longer L results in a longer life as the life of a sensor is measured in terms of the overall number of sleep-awake cycles until its energy is exhausted.

5   Concluding Remarks

In this work a protocol has been proposed which employs the asynchronous model originally presented in [13] and is lightweight in terms of the number of sleep/wake transitions and overall sensor awake time for training. Among the protocol variants, Flat− is the simplest one from the computational viewpoint because each sensor performs $O(1)$ operations per time slot. In contrast, Flat+ has the best performance for small values of d, but it cannot be used if the sensor is not allowed to skip one or more awake periods.

The results presented in this paper show that the protocol is flexible, in the sense that its parameters can be properly tuned. For instance, fixed the number k of coronas, one can decide the optimal values of d and L so as to minimize the number of sleep/wake transitions and/or the overall awake time per sensor. Conversely, one can fix the desired number of sleep/wake transitions, and then select suitable values of d and L.

However, several questions still remain open. First of all, it would be interesting to provide the analytical average behaviour of the protocol. In addition, a good idea for further work should be that of comparing the performance of the protocol proposed in the present paper with that devised in [3]. Indeed, the synchronous training protocol of [3] presents an irregular toggling between sleep and wake periods, so as to optimize the overall time for training, but it consumes energy in the explicit synchronization between the sensors and the sink to handle such irregular sleep/wake toggling. In contrast, the protocol proposed in Section 3 may force sensors to be awake for a longer time but avoids irregular toggling because sensors alternate between awake and sleep periods both of fixed length. Moreover, in this paper, a boolean (i.e. on/off) transmission model was assumed. That is, for each sink transmission range ir, all the sensors within a disk of radius ir around the sink hear the transmission, while all the sensors out of such a disk do not. Unfortunately, fading and shadowing impact on the connectivity of the network making such an on/off assumption almost impossible in practice [6]. Thus, the impact of the pseudo-ring corona in a real scenario is a very interesting aspect to be further investigated.

References

1. Akyildiz, I.F., Su, W., Sankarasubramanian, Y., Cayirci, E.: Wireless sensor networks: a survey. Computer Networks 38(4), 393–422 (2002)
2. Bandyopadhyay, S., Coyle, E.: An efficient hierarchical clustering algorithm for wireless sensor networks. In: Proc. IEEE INFOCOM 2003, San Francisco, CA, April 2003 (2003)
3. Bertossi, A.A., Olariu, S., Pinotti, M.C.: Efficient training of sensor networks. In: Nikoletseas, S.E., Rolim, J.D.P. (eds.) ALGOSENSORS 2006. LNCS, vol. 4240, pp. 1–12. Springer, Heidelberg (2006)
4. Culler, D., Estrin, D., Srivastava, M.: Overview of sensor networks. IEEE Computer 37(8), 41–49 (2004)
5. Ghiasi, S., Srivastava, A., Yang, X., Sarrafzadeh, M.: Optimal energy-aware clustering in sensor networks. Sensors 2, 258–269 (2002)
6. Gorce, J.M., Zhang, R., Parvery, H.: Impact of radio link unreliability on the connectivity of wireless sensor networks. EURASIP Journal on Wireless Communications and Networking (2007)
7. Griffin, H.: Elementary Theory of Numbers. McGraw Hill, New York (1954)
8. Langendoen, K., Reijers, N.: Distributed localization algorithm. In: Zurawski, R. (ed.) Embedded Systems Handbook, CRC Press, Boca Raton, FL (2004)
9. Lee, J.J., Krishnamachari, B., Jay, C.C.: Impact of heterogeneous deployment on lifetime sensing coverage in sensor networks. In: Proc. IEEE SECON (2004)
10. Nicolescu, D.: Positioning in ad-hoc sensor networks. IEEE Network 18(4), 24–29 (2004)
11. Olariu, S., Waada, A., Wilson, L., Eltoweissy, M.: Wireless sensor networks leveraging the virtual infrastructure. IEEE Network 18(4), 51–56 (2004)
12. Waada, A., Olariu, S., Wilson, L., Eltoweissy, M., Jones, K.: Training a wireless sensor network. Mobile Networks and Applications 10(1), 151–168 (2005)
13. Xu, Q., Ishak, R., Olariu, S., Salleh, S.: On asynchronous training in sensor networks. In: Proc. 3rd Intl. Conf. on Advances in Mobile Multimedia, K.Lumpur (September 2005)
14. Zhirnov, V.V., Herr, D.J.C.: New frontiers: self-assembly and nano-electronics. IEEE Computer 34(1), 34–43 (2001)

Optimal Placement of Ad-Hoc Devices Under a VCG-Style Routing Protocol*

Luzi Anderegg[1], Stephan Eidenbenz[2], Leon Peeters[1], and Peter Widmayer[1]

[1] Institute of Theoretical Computer Science, ETH Zurich
{anderegg,peeters,widmayer}@inf.ethz.ch
[2] Information Sciences (CCS-3), Los Alamos National Laboratory
eidenben@lanl.gov

Abstract. Motivated by a routing protocol with VCG-style side payments, this paper investigates the combinatorial problem of placing new devices in an ad-hoc network such that the resulting shortest path distances are minimum. Here, distances reflect transmission costs that are quadratic in Euclidean distance. We show that the general problem of placing multiple new wireless devices, either with different or identical transmission ranges, is NP-hard under multiple communication requests. On the positive side, we provide polynomial-time algorithms for the cases with only one new device and/or one communication request. To that end, we define geometric objects that capture the general geometric structure of wireless networks.

1 Introduction

Wireless ad-hoc networks promise the functionality of classical networks, without the burden of having to construct and install a fixed network infrastructure. Each wireless device in an ad-hoc network has a restricted transmission range, and communication between two devices typically takes place in a multi-hop fashion along intermediate devices. It is far from clear, however, if and why an intermediate device would be willing to sacrifice its own battery power and bandwidth to forward data packets destined for other devices.

Recently, several papers have addressed the issues caused by selfish devices in wireless ad-hoc networks. In particular, Anderegg and Eidenbenz [3] proposed a routing protocol that issues payments to the intermediate wireless devices, so as to compensate them for their energy costs. This compensation follows the marginal contribution principle by Vickrey, Clarke, and Groves, the key idea of the issued *VCG payments* being to reward a device for the gain in overall benefit that its participation causes (a good overview of VCG mechanisms is provided in [15]).

The VCG nature of the payments in [3] guarantees that devices will truthfully report their distances to other devices. In particular, the profit a device makes

---

* Supported by the Swiss National Science Foundation through the NCCR project Mobile Information and Communication Systems.

M. Kutyłowski et al. (Eds.): ALGOSENSORS 2007, LNCS 4837, pp. 58–70, 2008.

from the VCG payments depends on its position in the network. This motivates the main question studied in this paper: Where should a device position itself in the network so as to maximize its profit from the VCG payments? Although inspired by a game-theoretic setting, this is a purely combinatorial question that we study in the more general setting of several devices to be positioned for several communication requests, under a transmission cost model that is quadratic in Euclidean distance.

1.1 Model and Notation

More formally, we model the above setting as a graph $G = (V, E)$, with the vertex set $V = \{1, \ldots, n\}$ representing the set of incumbent wireless devices. Each vertex is embedded in the plane, and its coordinates are specified by a placement function $p : V \to \mathbb{R}^2$. We use the Euclidean distance measure, with $|uv|$ denoting the distance between two devices u and v, and also writing $|xx'|$ for the Euclidean distance between two points x, x' in the plane. We assume that the distance between any two device positions can be computed in constant time. The transmission ranges of the devices are modeled by a transmission range function $r : V \to \mathbb{R}_+$, specifying the maximal distance $r(u)$ from device u at which another device can still receive a signal from u via direct communication.

The edge set E of size m contains a directed edge (u, v) whenever device v lies within the transmission range of device u, that is, if and only if $|uv| \leq r(u)$. The cost $c(u, v)$ of a directed edge (u, v) reflects the energy requirement for transmitting a unit size data packet along the edge. Following the most common theoretical models of power attenuation, the cost is taken proportional to the squared Euclidean distance as $c(u, v) = \gamma |uv|^2$, with γ some constant. For convenience, we set the cost of all non-edges $(u, v) \notin E$ to $c(u, v) = \infty$.

The network needs to accommodate a number of communication requests between devices, which we model by a commodity set $K = \{(s_1, t_1), \ldots, (s_k, t_k)\}$, with s_i and t_i being the i-th source device and destination device, respectively. Each communication request is for a single unit size packet, and no two commodities share both the source and destination device. Hence, k can be as large as $\binom{n}{2}$. If there is only one commodity, then we denote the source by s and the destination by t. We refer to a tuple of the form (V, E, K, p, r, c) as a transmission graph T, where c depends on p and γ.

By $SP_T(s, t)$ we denote a shortest path in the transmission graph T from s to t with respect to the edge costs c. Further, $SP_T^{-u}(s, t)$ denotes the length of a shortest $s-t$-path not using the vertex u, and $SP_T^{-U}(s, t)$ the length of a shortest $s - t$-path not using any vertex in the set of vertices $U \subset V$. The total costs of any path P are denoted by $c(P)$, and by $c(T) = \sum_{i \in K} c(SP_T(s_i, t_i))$ we refer to the total path costs over all commodities (assuming that every commodity in K is connected by a path of finite cost).

For finding a shortest path from a source vertex s to a destination vertex t in a transmission graph T, [3] proposes the following incentive-compatible ad-hoc VCG protocol . First, the protocol basically asks the vertices for their positions and mutual distances during a flooding broadcast phase. Using this information, the protocol computes the edge costs $c(u, v)$, and a shortest path $SP_T(s, t)$. Finally, it pays each vertex $u \in SP_T(s, t)$ an amount $c(SP_T^{-u}(s, t)) - c(SP_T(s, t)) + c(u, v)$, where (u, v) is the outgoing edge of u in $SP_T(s, t)$. Because of the VCG nature of the payments, the computed $c(u, v)$ are equal to the true transmission costs. Thus, a vertex u gains a profit of $c(SP_T^{-u}(s, t)) - c(SP_T(s, t))$. This principle extends to the case where a selfish agent controls a set of devices U, and gains a profit of $c(SP_T^{-U}(s, t)) - c(SP_T(s, t))$.

1.2   The Device Placement Problem

Inspired by the ad-hoc VCG protocol, this paper takes the perspective of a profit maximizing selfish agent that enters an existing transmission graph T with a set $\Delta V = \{n+1, \ldots, n+\Delta n\}$ of Δn new devices, each with a maximal transmission range $r(v), v \in \Delta V$. Assuming that the communication requests for the near future are known, the agent's goal is to determine positions for its Δn devices such that the profit from the resulting VCG payments is maximum. Denoting by T' the new transmission graph including the new devices ΔV at their chosen positions, the objective function is defined as

$$\text{maximize} \sum_{i=1}^{k} \big(c(SP_T(s_i, t_i)) - c(SP_{T'}(s_i, t_i)) \big) = c(T) - c(T'). \qquad (1)$$

Since the first term in Equation 1 is independent of the positions of the devices in ΔV, the problem is equivalent to:

$$\text{minimize } c(T'). \qquad (2)$$

Thus, besides the game-theoretic motivation, the resulting *device placement problem* can also be defined in purely combinatorial terms: place Δn additional devices such that the shortest paths in the resulting transmission graph T' are of minimum length.

We investigate the algorithmic complexity of the device placement problem for $\Delta n = 1$ as well as for general Δn, and for $k = 1$ communication request as well as for general k. Depending on the form of the maximal transmission ranges of the additional devices, we study two problem variants: with *identical* new devices that each have the same transmission range $r(n + 1) = r(n + 2) = \ldots = r(n + \Delta n)$, and with *individual* new devices, each having their own transmission range $r(v), v \in \Delta V$. Clearly, the two problem variants do not differ for a single additional device. Hence, we simply refer to this case as the single device placement problem.

1.3   Related Work

Network upgrade problems where an existing network has to be extended such that the resulting network exhibits certain properties are classical optimization problems. Several variants of these problems have been considered, and the work closest to ours, although still quite different, is the thesis by Krumke [14]. Given a graph and a function specifying the cost of shortening an edge, he investigated how to determine an optimal strategy to minimize the total weight of a minimum spanning tree within a budget restriction.

The idea of our approach in this paper is in a similar vein to the work on network creation games in [10]. The main goal there is to explain the structure of networks constructed by independent selfish agents from a game-theoretic point of view. Different authors [2,5,6,7,12] continued this line of study in related network creation models.

1.4 Our Contribution

We show that the most general problem of placing Δn new wireless devices, each with its own transmission range, is already NP-hard for only $k = 2$ communication requests. We also prove that the problem is still NP-hard when the newly placeable devices have identical transmission ranges, but in this case for a general number k of communication requests.

On the positive side, we provide a polynomial-time algorithm for the problem of placing Δn new devices with identical transmission ranges under a single communication request. To arrive at this result, we first study the case with a single communication request and a single new device, analyze its geometric structure, and propose geometric objects that capture this structure. We further present a polynomial-time algorithm for optimally placing a single new device under k communication requests.

2 Placing Multiple Devices for Multiple Commodities

We first show that the problem of placing multiple identical new devices for multiple commodities is NP-hard. The decision version of this problem is stated as follows:

PROBLEM: IDENTICAL DEVICE PLACEMENT.
INSTANCE: An instance $I = (T, \Delta n, r, Z)$ consists of a transmission graph $T = (V, E, K, p, r, c)$, a positive integer Δn, an identical maximal transmission range $r(v)$ for each additional device $v \in \Delta V$, and a positive number Z.
QUESTION: Is there a placement for the Δn additional devices such that the difference $c(T) - c(T') \geq Z$, where T' is the transmission graph after the placement of the additional devices?

Theorem 1. IDENTICAL DEVICE PLACEMENT *is NP-hard.*

As the proof of Theorem 1 is quite lengthy and involved, we only present a proof sketch. The entire proof can be found in the appendix of the full version of this paper [4].

Proof sketch. The main idea of the proof is to reduce a restriction of planar EXACT COVER BY 3-SETS (X3C) to IDENTICAL DEVICE PLACEMENT. In an

X3C instance $I(U, S, b)$ we are given a set U of $3b$ elements, a collection of 3-element subsets $S = \{S_1, \ldots, S_{|S|}\}$ of U, and a budget b. We are looking for a subcollection of size b from S whose union is U. We use the restricted version of X3C where the corresponding bipartite graph is planar and each element appears in either two or three sets [9], and denote it by X3C-3.

The idea of the reduction is to introduce a device for every element and every set from the X3C-3 instance. For embedding these devices in the plane, we use a result by Kant [13] for drawing a triconnected 3-planar graph with horizontal and vertical edge segments on a grid. Every element device forms one source-destination pair with an additional global destination device.

Further, we show that one can construct a *chain* consisting of a polynomial number of devices at and between any two points x, x', such that the cost of the shortest path between the devices at x and x' is bounded by their distance. Using such chains, we ensure that the only possible paths between an element device and the global destination device go through the set devices the element is member of. The cost of these paths is the same for all source-destination pairs. Moreover, the placement of an additional device within a chain yields only a small profit. Indeed, only one position induces a large improvement between each set device and the global destination device. Hence, the number of reasonable positions for the additional devices is limited to the number of subsets in S, and there is a one-to-one correspondence between such a position and a subset.   □

For individual devices that each have a specific maximal transmission range, we have to specify exactly which additional device is placed at which position. The decision version of the corresponding problem is defined as follows.

PROBLEM: INDIVIDUAL DEVICE PLACEMENT
INSTANCE: An instance $I = (T, \Delta n, r, Z)$ consists of a transmission graph $T = (V, E, K, p, r, c)$, a positive integer Δn, an individual maximal transmission range $r(v)$ for each additional device $v \in \Delta V$, and a positive number Z.
QUESTION: Is there a placement for the Δn additional devices such that $c(T') \leq Z$, where T' is the transmission graph after the placement of the additional devices?

Since the identical device placement problem is a special case of the individual device placement problem, it immediately follows that INDIVIDUAL DEVICE PLACEMENT is NP-hard. Below, we prove that it is already NP-hard for only two commodities.

Theorem 2. INDIVIDUAL DEVICE PLACEMENT *is NP-hard for* $k = 2$ *commodities.*

Proof. The proof is by a reduction from PARTITION (SP12 in [11]). In PARTITION, we are given a set $A = \{a_1, \ldots, a_{|A|}\}$ of positive integer numbers. The goal is to decide whether there is a subset $A' \subseteq A$ such that $\sum_{a_i \in A'} a_i = B/2$, where $B = \sum_{a_i \in A} a_i$. We construct a device placement instance consisting of the four devices $\{1, \ldots, 4\}$, placed at the vertices of a square with device 1 at position $\langle 0, 0 \rangle$, device 2 at position $\langle B/2 + 1, 0 \rangle$, device 3 at position $\langle 0, M \rangle$, and device 4 at position $\langle B/2 + 1, M \rangle$, with M an integer much larger than B. The maximal transmission ranges of these devices are set to one. Further, the device pair $(1, 2)$ constitutes the first commodity, and the device pair $(3, 4)$ the other one. The number Δn of additional devices is set to $|A|$, and the maximal transmission range $r(u)$ is set to a_{v-n}, for $v = n+1, \ldots, n + \Delta n$. Finally, we set Z to $2 + \sum_{a_i \in A} a_i^2$. A solution of the partition problem immediately gives a solution for the device placement problem: we place the devices with index in A' one after another on the line segment between s_1 and t_1, starting at distance 1 from s_1, such that their maximal transmission ranges are just exactly large enough to reach the next device. The remaining devices in $A \setminus A'$ are placed between s_2 and t_2 in a similar fashion. The total cost of the shortest path for the first commodity is now equal to $1 + \sum_{a_i \in A'} a_i^2$, and that for the second commodity is equal to $1 + \sum_{a_j \in A \setminus A'} a_j^2$. If the partition problem has no solution, then no placement of the devices connects both source-destination pairs, and total shortest path cost of infinity cannot be avoided. $\qquad\square$

3   Identical Device Placement for a Single Commodity

This section studies the basic geometric structure of the identical device placement problem for a single commodity. We first characterize the optimal position of one additional device. Next, we use that characterization to construct an algorithm for optimally placing an additional device, and extend that algorithm to compute the optimal positions of multiple identical devices.

3.1   The Optimal Position of an Additional Device

Suppose the transmission graph consists of only two devices u and v that wish to communicate, and we are interested in the best position for an additional device v'. Let the impact $F_{uv}(v')$ of the additional device v' be the difference between the cost of the direct communication from u to v and the cost of the communication from u to v via the additional device v'. That is, $F_{uv}(v') = c(u, v) - (c(u, v') + c(v', v))$. Note that the impact may be negative. Figure 1 illustrates the following observation relating the impact $F_{uv}(v')$ to the position of v'.

Observation 1. *The impact of an additional device v' between devices u, v is equal to $F_{uv}(v') = c(u, v)/2 - 2\gamma \cdot |v', M_{uv}|^2$, where M_{uv} is the middle point of the line segment from u to v.*

Observation 1 implies that device positions with the same impact lie on a circle with center M_{uv}, with the maximum impact achieved at M_{uv}. From there the impact decreases quadratically in each direction, and it is equal to zero for positions on a circle with center M_{uv} and radius $|uv|/2$.

Next, we include a single source-destination pair (s, t) into the impact function. To that end, we define the impact $F_{uv}^{st}(v')$ of an additional device v' on a

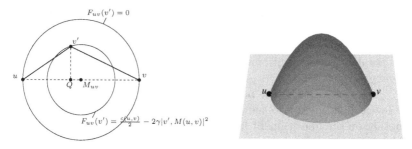

Fig. 1. Points with same impact $F_{uv}(v')$ are on circles around $M(u,v)$ (left), and the corresponding graph of $F_{uv}(v')$ in $\mathbb{R}^3$

device pair (u,v) with respect to the single source-destination pair (s,t) as the difference between the shortest $s - t$-path length without v', and with v' and $(u,v'),(v',v)$ as a *mandatory* partial path.

Observation 2. *The impact of an additional device v' for a device pair (u,v) and a source-destination pair (s,t) is:*

$$F_{uv}^{st}(v') = c(SP_T(s,t)) - c(SP_T(s,u)) - 2\gamma \cdot |v', M_{uv}|^2 - c(u,v)/2 - c(SP_T(v,t)).$$

Proof. $F_{uv}^{st}(v') = c(SP_T(s,t)) - [c(SP_T(s,u)) + c(u,v') + c(v',v) + c(SP_T(v,t))]$, so the observation follows by using Observation 1. □

Note that the impact is defined for every pair of devices $u, v \in V$, and that it can be negative. An additional device induces a shortest path along (u,v',v) if its impact is positive. The impact is again equal for all positions with the same distance to M_{uv}, and the maximum impact is achieved at position M_{uv}. Observe that the circle with positions of zero impact does not necessarily go through the positions of devices u and v. Indeed, if u and v are not on a shortest path before inserting the additional device, then the circle with positions of zero impact has a smaller radius than $|M_{uv}, u|$.

Some positions with positive impact may be unreachable due to small maximal transmission ranges of both the additional device and the existing devices. Thus, we define the profit region PR_{uv}^{st} as the set of positions for an additional device v' where $F_{uv}^{st}(v')$ is positive, u can reach v', and v' can reach v, given the maximal transmission ranges. Geometrically, a profit region PR_{uv}^{st} is the intersection of three disks: the disk around M_{uv} where $F_{uv}^{st}(v') \geq 0$, the disk with center u and radius $r(u)$, and the disk with center v and radius $r(v')$. See Figure 2 for three possible shapes of such an intersection. The boundary of a shape consists of at most four circle segments. We define $G_{uv}^{st}(\cdot)$ to be the function $F_{uv}^{st}(\cdot)$ restricted to the corresponding profit region. That is, $G_{uv}^{st}(v')$ is equal to $F_{uv}^{st}(v')$ for all positions of v' inside PR_{uv}^{st}, and $-\infty$ otherwise.

Assuming that the profit region PR_{uv}^{st} is not empty and that v' is placed between u and v, the position inside PR_{uv}^{st} with minimal distance to the point M_{uv}

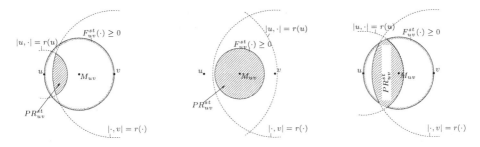

Fig. 2. Profit region $PR(u, v, (s,t))$ building an asymmetric lens, a circle, and a shape bounded by four arcs

is the best position for v' because it maximizes $G_{uv}^{st}(v')$. If the maximal transmission ranges of u and v' are large enough, then this is the same as M_{uv} itself. If the maximal transmission range of v' or u is too small, then the best position moves on the line segment between u and v towards device u respectively v until it enters the profit region. Such a best position is denoted by $p^*(u, v, (s,t))$, and it can be computed in constant time given the distance between u and v and the maximal transmission ranges $r(u)$ and $r(v')$.

3.2 Multiple Identical Device Placement for a Single Commodity

The fact that one additional device reduces the cost between exactly one device pair enables us to state the following geometric formulation for placing a single additional device for a single commodity:

$$\max_{p(v') \in \mathbb{R}^2} \max_{u,v \in V} G_{uv}^{st}(v'). \tag{3}$$

Below, we use this formulation to derive an algorithm for placing multiple identical devices for a single commodity. As a first step, however, we note that Observation 2 and formulation (2) together induce an algorithm for the simpler problem of optimally placing a single device for a single commodity.

To that end, we define the following *expanded 2-layer graph* to encode the restriction that only one additional device is available. The graph has two layers, labeled 0 and 1, each containing a copy of the transmission graph. We add an edge from each vertex $(u, 0)$ on layer 0 to each vertex $(v, 1)$ on layer 1, for $u \neq v$. The cost of such an edge is equal to $c(u, p^*) + c(p^*, v)$, the transmission cost between u and v via an additional device at position $p^*(u, v, (s,t))$. For simplicity, we exclude edges with infinite cost. In this graph, we then search a shortest path from vertex $(s, 0)$ to vertex $(t, 0)$ and another one from $(s, 0)$ to vertex $(t, 1)$. By construction, the minimum of these two paths corresponds to optimally placing the additional device. Note that a shortest path does not necessarily use the additional device as the maximal transmission range of the additional device might be too small to be useful.

The above approach can be extended as follows to optimally place Δn *identical* additional devices, instead of only one. In principle, the best positions for

$h \leq \Delta n$ additional devices between a fixed pair (u, v) of devices are to distribute the additional devices in equal distances on the segment connecting u and v. However, limited maximal transmission ranges of device u or of the additional devices may make such equal distances impossible. In such a case, the additional devices are distributed on the feasible part of the segment connecting u and v as evenly as possible. Based on this insight, we construct a $(\Delta n + 1)$-layer graph $H = (V_H, E_H)$ with a copy of the transmission graph on each layer. For each layer $h < \Delta n$ and each 'higher' layer $h' > h$, we add an edge from each vertex (u, h) to each vertex (v, h'), for $u \neq v$, the cost of which are equal to the transmission cost from u to v via $(h' - h)$ optimally placed additional devices between u and v, as discussed before. Edges with infinite cost are again excluded for simplicity.

Theorem 3. *The multiple identical device placement problem for a single commodity can be solved in time* $O((\Delta n)^2 n^2)$.

Proof. We use the $(\Delta n + 1)$-layer graph H described above, compute a shortest path between $(s, 0)$ and (t, h) for each $h, 0 \leq h \leq \Delta n$, and output a path with length $\min_{0 \leq h \leq \Delta n} c(SP((s, 0), (t, h)))$. Since the cost of each edge $((u, h), (v, h')) \in E_H$ correctly reflects the cost of a subpath from u to v containing exactly $(h' - h)$ optimally placed additional devices between u and v, the correctness of the algorithm follows. The construction of H needs time $O((\Delta n)^2 n^2)$, as there are that many potential edges in the graph. All shortest paths can be found in time $O((\Delta n)^2 n^2)$ using Dijkstra's algorithm to find a shortest path tree rooted at $(s, 0)$.   □

4   Single Device Placement for Multiple Commodities

With multiple commodities($k > 1$), the optimal position for a single additional device may be different from the optimal point $p^*(u, v, (s_i, t_i))$ between some existing devices u and v, and a specific commodity i. Rather, the best position could be a position where connections between several source-destination pairs use the new device. Unfortunately, the ideas from the previous section do not easily extend to a polynomial-time algorithm for the single device and multiple commodities case. Therefore, we first present a different algorithm for the single device and single commodity case, which has worse running time than the algorithm above, but is extendable to the single device and multiple commodities case.

4.1   Single Maximization Diagram Approach

An alternative approach to solve the single device and single commodity case is to directly use the geometric formulation in (3). There, the term $\max_{u,v \in V} G_{uv}^{st}(\cdot)$ defines exactly the *upper envelope* of the impact functions $G_{uv}^{st}(\cdot), u, v \in V$, that is, the the point-wise maximum of the curves $G_{uv}^{st}(\cdot)$. The *maximization diagram* $\mathcal{M}$ of the impact functions $G_{uv}^{st}(\cdot)$ divides the plane into maximal connected

Fig. 3. Upper envelope of the impact functions for two device pairs, and the corresponding maximization diagram

cells, such that within one cell the same function $G_{uv}^{st}(\cdot)$ attains the upper envelope defining maximum (see [1] for a detailed description of maximization diagrams). Figure 3 shows the upper envelope of two impact functions, and the corresponding maximization diagram. Thus, a cell in $\mathcal{M}$ has a *characterizing* device pair, and for each point in the cell, that device pair yields the maximum impact. Inside a given cell, the optimal position for a new device is defined by the maximum of the concave function $G_{uv}^{st}(\cdot)$ for the characterizing device pair (u, v), and is hence easy to compute. For a polynomially bounded number of cells, this approach gives rise to a polynomial time algorithm.

Figure 3 illustrates that an edge in $\mathcal{M}$ arises either from the intersection of two impact functions $G_{uv}^{st}(\cdot)$, or from a domain boundary of an impact function. These domain boundaries are circle segments, and the following observation states that an intersection yields a line segment. Thus, the edges of any maximization diagram cell are either line or circle segments.

Observation 3. *The intersection of two impact functions $G_{u_1 v_1}^{st}(\cdot)$ and $G_{u_2 v_2}^{st}(\cdot)$ is a line.*

Proof. Consider the impact functions for device pairs (u_1, v_1) and (u_2, v_2), and an additional device v'. Both impact functions $G_{u_i v_i}^{st}(v')$ are of the form $H_i - 2 \cdot \gamma |v', M_{u_i v_i}|$, where the constant H_i depends on the positions of the devices, for $i = 1, 2$. If we set $G_{u_1 v_1}^{st}(\cdot) = G_{u_2 v_2}^{st}(\cdot)$, then the set of points fulfilling the equation constitutes a line. $\square$

Lemma 1. *Given a 2-dimensional maximization diagram cell c, represented by a list of its n_c incident edges, with a characterizing pair (u, v), the optimal position inside c with respect to $G_{uv}^{st}(\cdot)$ can be found in time $O(n_c)$.*

Proof. Inside c, the profit of any position is equal to the concave impact function $G_{uv}^{st}(\cdot)$. Hence, the maximum inside c is either attained at the single point where the gradient is equal to zero, if this point lies inside c, or it is attained somewhere on the boundary of c. For the function $G_{uv}^{st}(\cdot)$, the gradient is zero at position M_{uv}, and if this position is inside c, we are done. Otherwise, we go along the

---

Algorithm. MaxDiagram(s, t).

Output. Optimal position for one additional device for one commodity (s, t).

forall device pairs (u, v) **do**
  $\lfloor$ compute $G_{uv}^{st}(\cdot)$;
compute maximization diagram $\mathcal{M}$ of $\cup_{u,v} G_{uv}^{st}(\cdot)$;
compute the global optimum over all 2-dimensional max. diagram cells $c \in \mathcal{M}$;

---

boundary edges of c, where, for a single edge, the maximum is attained at the position with smallest distance to M_{uv}.

Testing whether M_{uv} is inside c can be done in time linear in n_c by comparing the position to each edge. The maximum computation for all n_c edges needs linear time as well, since the position on a line segment or circle segment edge with smallest distance to M_{uv} can be determined in constant time. $\square$

Lemma 2. *The single device placement problem for a single commodity can be solved in time $O(n^{4+\epsilon})$.*

Proof. We use Algorithm MaxDiagram that extends the above approach by considering all maximization diagram cells. First, we compute the single source shortest paths tree from s, and the single destination shortest paths tree to t. This can be done in time $O(n \log n + m)$. The for-loop over all device pairs needs time $O(n^2)$, and within one iteration we evaluate G_{uv}^{st} for a device pair (u, v). As a single evaluation can be executed in constant time using the shortest path trees, this step runs in time $O(n^2)$.

It was shown in [1] that the maximization diagram of ℓ partially defined functions in $\mathbb{R}^3$ can be computed in time $O(\ell^{2+\epsilon'})$, for any $\epsilon' > 0$. Thus, the maximization diagram $\mathcal{M}$ of the $O(n^2)$ functions $G_{uv}^{st}(\cdot)$ can be computed in time $O(n^{4+\epsilon})$, and the combinatorial complexity of $\mathcal{M}$ is $O(n^{4+\epsilon})$ as well. Using Lemma 1 and the fact that each edge is incident to at most two cells, computing the maximum over all 2-dimensional cells in $\mathcal{M}$ takes time $O(n^{4+\epsilon})$. All together, the running time is $O(n^{4+\epsilon})$. $\square$

4.2   Multiple Maximization Diagrams Approach

Next, we extend the above approach to the single device placement problem for multiple commodities, by means of the following geometric formulation:

$$\max_{p(v') \in \mathbb{R}^2} \sum_{i=1}^{k} \max_{u,v \in V} G_{uv}^{s_i t_i}(v'). \tag{4}$$

We first compute the maximization diagram $\mathcal{M}_i$ for each commodity $i \in K$. Now, each point in the plane is part of one cell in each $\mathcal{M}_i$, and that cell determines the characterizing pair for the corresponding commodity i (if it exists). We use this fact to determine the regions in which each point has the same characterizing

---

Algorithm. MaxDiagramOverlay(K).

Output. Optimal position for one additional device for the commodity set K.

forall commodities $i \in K$ **do**

    forall device pairs (u, v) **do**

        compute $G_{uv}^{s_i t_i}(\cdot)$;

    compute maximization diagram $\mathcal{M}_i$ of $\cup_{u,v} G_{uv}^{s_i t_i}(\cdot)$;

compute $\mathcal{O} = overlay(\mathcal{M}_1, \ldots, \mathcal{M}_k)$;

compute the global optimum over all 2-dimensional overlay cells $c \in O$;

---

pair *for each single commodity*. More precisely, we construct the *overlay* $\mathcal{O}$ of the cell sets $\mathcal{M}_1, \ldots, \mathcal{M}_k$ (see Chapter 2 in [8] for a discussion of overlays). Then, for every single commodity, the characterizing pair is the same for every point in an overlay cell. Lemma 3 states the complexity of computing the optimal position inside a single overlay cell, and Theorem 4 the resulting complexity of the above approach.

Lemma 3. *Given a 2-dimensional overlay cell c with a (possibly empty) characterizing pair (u_i, v_i) for each commodity $i \in K$, and represented by a list of its n_c incident edges, the optimal position inside c with respect to the profit $\sum_i G_{uv}^{s_i t_i}(\cdot)$ can be found in time $O(n_c)$.*

Proof. The position for which $\sum_i G_{uv}^{s_i t_i}(\cdot)$ attains the maximum is the optimal position inside c. As the functions $G_{uv}^{s_i t_i}(\cdot)$ are concave inside c for all characterizing pairs, and for all commodities $i \in K$, the sum over these functions is concave as well. As in Lemma 1, the maximum of the resulting concave function is either attained at the single point where the gradient is zero, or on the boundary of c. Here, the single point where the gradient is zero evaluates to the center of mass of the positions $M_{u_i v_i}$. The remainder of the proof is the same as in the proof of Lemma 1. □

Theorem 4. *The single device placement problem for multiple commodities can be solved in time $O(k^2 n^{8+2\epsilon} \log (kn^{4+\epsilon}))$.*

Proof. We use Algorithm MaxDiagramOverlay that summarizes the above described approach. The nested for-loop needs time $O(kn^{4+\epsilon})$ as we compute a maximization diagram for each of k commodities. The overlay of two sets of planar geometric objects with combinatorial complexities ℓ' and ℓ'' can be computed in time $O(\ell \log (\ell' + \ell''))$ where ℓ is the combinatorial complexity of the resulting overlay (see Chapter 2 in [8]). As the combinatorial complexity of the overlay is $O((kn^{4+\epsilon})^2)$, it can be constructed in time $O(k^2 n^{8+2\epsilon} \log (kn^{4+\epsilon}))$. Using Lemma 3 and the fact that each edge is incident to at most two cells, the running time for computing the global optimum over all overlay cells is in $O(k^2 n^{8+\epsilon})$. Thus, the computation of the overlay dominates the overall running time of the algorithm. □

References

1. Agarwal, P., Schwarzkopf, O., Sharir, M.: The overlay of lower envelopes and its applications. Discrete & Computational Geometry 15(1), 1–13 (1996)
2. Albers, S., Eilts, S., Even-Dar, E., Mansour, Y., Roditty, L.: On Nash equilibria for a network creation game. In: Proceedings SODA 2006, pp. 89–98 (2006)
3. Anderegg, L., Eidenbenz, S.: Ad hoc-VCG: a truthful and cost-efficient routing protocol for mobile ad hoc networks with selfish agents. In: Proceedings MOBICOM 2003, pp. 245–259 (2003)
4. Anderegg, L., Eidenbenz, S., Peeters, L., Widmayer, P.: Optimal placement of ad-hoc devices under a VCG-style routing protocol (2007),
   http://www.inf.ethz.ch/personal/peetersl/publications.html
5. Anshelevich, E., Dasgupta, A., Kleinberg, J., Tardos, E., Wexler, T., Roughgarden, T.: The price of stability for network design with fair cost allocation. In: Proceedings FOCS 2004, pp. 295–304 (2004)
6. Anshelevich, E., Dasgupta, A., Tardos, E., Wexler, T.: Near-optimal network design with selfish agents. In: Proceedings STOC 2003, pp. 511–520 (2003)
7. Corbo, J., Parkes, D.: The price of selfish behavior in bilateral network formation. In: Proceedings PODC 2005, pp. 99–107 (2005)
8. de Berg, M., van Kreveld, M., Overmars, M., Schwarzkopf, O.: Computational Geometry: Algorithms and Applications. Springer, Heidelberg (1997)
9. Dyer, M., Frieze, A.: Planar 3DM is np-complete. Journal of Algorithms 7, 174–184 (1986)
10. Fabrikant, A., Luthra, A., Maneva, E., Papadimitriou, C., Shenker, S.: On a network creation game. In: Proceedings PODC 2003 (2003)
11. Garey, M., Johnson, D.: Computers and Intractability: A Guide to the Theory of NP-Completeness. W. H. Freeman & Co, New York (1979)
12. Hayrapetyan, A., Tardos, E., Wexler, T.: A network pricing game for selfish traffic. In: Proceedings PODC 2005, pp. 284–291 (2005)
13. Kant, G.: Drawing planar graphs using the lmc-ordering. In: Proceedings FOCS 1992, pp. 101–110 (1992)
14. Krumke, S.: The approximability of location and network design problems. PhD thesis, University of Wuerzburg (1996)
15. Mas-Colell, A., Whinston, M., Green, J.: Microeconomic Theory. Oxford University Press, New York (1995)

Analysis of the Bounded-Hops Converge-Cast Distributed Protocol in Ad-Hoc Networks*

Marcin Zawada

Institute of Mathematics and Computer Science
Wroclaw University of Technology
Poland
marcin.zawada@pwr.wroc.pl

Abstract. We consider the problem of bounded hops converge cast in ad-hoc networks. Let us assume that stations are located on the d-dimensional Euclidean space and there is one distinguished station called a base station. This problem, called the d-DIM h-HOPS CONVERGECAST, is defined as finding a minimal energy-cost range assignment, which allows each station to communicate with a base station in at most h hops. Clementi *et al.* [2] proposed a distributed protocol h-PROT for $d = 2$ and proved that in case of $h = 2$ the expected approximation ratio of this protocol is $O(1)$ on random instances. However, for $h = 3, \ldots, 8$ they provided only an experimental study showing that the protocol has good performances. In this paper, we introduce the protocol (d, h)-PROT which extends the protocol h-PROT on the d-dimensional space. We address the probabilistic analysis and show formally that the protocol (d, h)-PROT achieves an approximation ratio of $O(1)$ in expectation on random instances for any $d, h \geq 2$.

1   Introduction

We consider an ad-hoc network consisting of processing units, called *stations* that are located on the d-dimensional Euclidean space. We assume that S denotes a set of stations and we assign to each station a *transmission range* $\mathcal{R} \colon S \to \mathbb{R}^+$. The transmission range of a station $s \in S$ is exactly the area in which another station $t \in S$ can receive messages sent by s, i.e. $d(s, t) \leq \mathcal{R}(s)$, where $d(s, t)$ is the Euclidean distance between s and t. The transmission range depends on the energy power supplied to the station and we assume that stations can change their transmission ranges by adequately supplying the energy power. The power P_s, required by a station s to exchange data with another station t, must satisfy the inequality

$$P_s > \gamma \cdot d(s, t)^\alpha, \tag{1}$$

where $\alpha \geq 1$ is the *path loss exponent*, and $\gamma \geq 1$ is the *transmission quality* parameter. The parameter α depends on the environment conditions and a terrain structure, and can vary between 1 and more then 6 in heavily built urban

---

* Partially supported by the EU within the 6th Framework Programme under contract 001907 (DELIS).

M. Kutyłowski et al. (Eds.): ALGOSENSORS 2007, LNCS 4837, pp. 71–82, 2008.

areas. The $\alpha = 2$ for propagation in free space. Let $S = \{s_1, s_2, \ldots, s_n\}$. Notice that a range assignment $\mathcal{R}$ determines a directed *transmission graph* $\mathcal{G}(S, E)$, where edge $(s_i, s_j) \in E$ if and only if $d(s_i, s_j) \leq \mathcal{R}(s_i)$. We need to find such a range assignment $\mathcal{R}$ for which the corresponding transmission graph $\mathcal{G}$ satisfies connectivity constraints and the overall energy supplied is minimized. Thus the overall energy (i.e. the *cost*) of a range assignment $\mathcal{R} \colon S \to \mathbb{R}^+$ is defined as

$$cost(\mathcal{R}) = \sum_{s \in S} \mathcal{R}(s)^\alpha. \qquad (2)$$

In this paper, we address the range assignment problem in which $\mathcal{G}$ is required to contain a tree directed towards a given *base station* b, spanning S and of depth at most h.

The d-DIM h-HOPS CONVERGECAST problem is a particular case of the well-know *Minimal h-hops Spanning Tree* problem (h-HOPS MST) which is defined as follows. Let $\mathcal{G}(V, E)$ be a graph with non-negative edge weights and $b \in V$. We need to find a minimum-cost directed tree rooted at b of depth at most h and spanning the graph $\mathcal{G}$. In fact, d-DIM h-HOPS CONVERGECAST corresponds to h-HOPS MST, where nodes represent stations and edges represent communication links between stations in such a way, that for any pair of stations s_i and s_j there exists an edge if their weight is $d(s_i, s_j)^\alpha$.

Clementi *et al.* proposed in [2] an efficient distributed heuristics h-PROT for 2-DIM h-HOPS CONVERGECAST and analyzed its expected solution cost for $h = 2$ hops. They proved that for $h = 2$ their protocol has a constant expected approximation ratio and stated an open problem whether the expected approximation ratio for any $h > 2$ is constant or not. In this paper, we address this problem. By Theorem 4 for $d = 2$, we show that the expected approximation ratio of the protocol h-PROT is constant for any $h \geq 2$. Furthermore, we generalize the protocol for the d-DIM h-HOPS CONVERGECAST problem and prove that it has an approximation ratio of $O(1)$ in expectation for any $d, h \geq 2$.

The structure of this paper is as follows. In Section 2 we give an overview of previous results. The algorithm (d, h)-PROT is introduced in Section 3. In Section 4 we provide probabilistic analysis of the algorithm and we prove that the expected energy cost in the network is within $O(1)$ of an optimal solution. We conclude in Section 5.

2   Previous Results

Alfandari and Paschos [1] proved that even the 2-HOPS MST with non-negative edge weights is NP-hard and that the metric version of the problem is MaxSNP-hard. Moreover, they proved that the general 2-HOPS MST in graph of order n cannot be approximated within better then $O(\log n)$ of optimum. As pointed out by Gouveia in [9], the 2-DIM 2-HOPS MST is equivalent to the *Simple Uncapacitated Facility Location* (SUFL) problem. Thus, we can apply existing approximation algorithms for SUFL to 2-DIM 2-HOPS MST. Furthermore, in [4,9] authors evaluated and compared solutions for the d-DIM h-HOPS MST problem on random instances, mainly by performing computer experiments. However, just recently Clementi et al. [3,5] showed a tight analysis of the expected optimal cost for the d-DIM h-HOPS MST on random instances. They proved the following theorem.

Theorem 1 ([3,5]). *Let h and d be fixed positive integers. Let S be a random instance of n points in a d-cube of side length L and let T be any tree of height h and spanning S. Then, it holds that*

$$cost(T) = \begin{cases} \Theta\left(L \cdot n^{\frac{1}{h}}\right), & if\ d = 1, \alpha = 1, \\ \Theta\left(L \cdot n^{1-\frac{1}{d}+\frac{d-1}{d^{h+1}-d}}\right), & if\ d \geq 2, \alpha = 1, \\ \Theta\left(L^2 \cdot n^{\frac{1}{h}}\right), & if\ d = 2, \alpha = 2 \end{cases}$$

with probability at least $1 - e^{-c \cdot n}$, for some constant $c > 0$.

Moreover, they presented the asymptotically cost-optimal heuristic h-PARTY for the d-DIM h-HOPS MST problem on random instances. The proposed heuristic is based on the well-know technique *divide and conquer* and requires global knowledge of the network and centralized decisions. They showed that the centralized heuristic h-PARTY has the approximation ratio of $O(1)$. However, the key issue of the optimality of the heuristic h-PARTY is the size of the grid partition of the d-cube. Thus, in a distributed model this heuristic would be very expensive and therefore impractical.

To that end, in [2] Clementi et al. proposed a distributed protocol h-PROT to address the 2-DIM h-HOPS CONVERGECAST problem. They analyzed the expected solution costs for $h = 2$ hops and proved the following.

Theorem 2 ([2]). *Let S be a random instance of n nodes selected from a square Q of edge size L and let $b \in S$ be a base station. Then, for $h = 2$, the expected cost of the range assignment $\mathcal{R}$ returned by protocol h-PROT satisfies the following bounds*

$$\mathbb{E}[cost(\mathcal{R})] = O(L^{\alpha} n^{\frac{2}{\alpha+2}}) .$$

As a corollary of Theorem 2, they obtained that the protocol 2-PROT has the expected approximation ratio of $O(1)$. They also stated as an open problem, whether the expected approximation ratio for any $h > 2$ is $O(1)$ or not.

3 Distributed Protocol (d, h)-Prot

We consider d-DIM h-HOPS CONVERGECAST on an ad-hoc network and a distributed randomized protocol (d, h)-PROT (see Fig. 1) which, for a given set S of stations and a base station $b \in S$, constructs a feasible range assignment $\mathcal{R}$. The distributed protocol (d, h)-PROT is a quite straightforward extension of the existing protocol h-PROT (see [2]) on a d-dimensional space. However, beside adding modifications of the way messages have to be sent on greater distances (line *2* and *11*, Fig. 1), we properly modify the probability $p^*(d, h, j, n, \alpha)$ (line

procedure (d, h)-PROT(b,n,α,L)

```
var     level       : integer
        connected   : boolean
        coin        : integer
```

1.   **if** $s = b$ **then**
2.   **send**("start",$\sqrt{d}L$); *connected* := **true**; *level* := 0;
3.   **else**
4.   **wait**("start"); *connected* := **false**;
5.   **for** $j = 1, \ldots, h$ **do**
6.   **if** *connected* = **false then**
7.   **if** $j < h$ **then**
8. $coin := \begin{cases} 1, & \text{with probability } p^*(d,h,n,j-1,\alpha) \\ 0, & \text{with probability } 1 - p^*(d,h,n,j-1,\alpha) \end{cases}$
9.   **else**
10. $coin := 1$
11.   **for** $r = 2^0, 2^1, \ldots, 2^{\log\lceil\sqrt{d}L\rceil}$ **do**
12.   **if** *connected* = **false** and *coin* = 1 **then**
13.   **if received**("echo:v") = **true then**
14. $\mathcal{R}(s) := d(s, v)$; *connected* := **true**; *level* := j;
15.   **else**
16.   **send**("search:s", r);
17.   **else**
18.   **if** *level* = $j - 1$ **then**
19.   **if received**("search:v") = **true then**
20.   **if** $d(s, v) = \min\{d(s', v) : s'$ **received** "search:v"$\}$ **then**
21.   **send**("echo:s", $d(s, v)$);

Fig. 1. Pseudo-code of (d, h)-PROT algorithm executed by a single station $s \in S$

8, Fig. 1) on the d-dimensional space. The important and the most difficult part is to formally analyse the protocol. In Section 4, we provide a formal probabilistic analysis and prove that the protocol (d, h)-PROT has indeed the expected approximation ratio of $O(1)$, for any $d, h \geq 2$.

Firstly, we give a brief description of the main idea of the protocol (d, h)-PROT. As we mentioned before, apart some necessary modifications, the protocol (d, h)-PROT is based on the existing protocol h-PROT, so its description is similar to the one we can find in [2]. Therefore, we only briefly outline how the protocol constructs a range assignment $\mathcal{R}$. We assume that every station knows only its label, position on the d-cube and its side length L. If stations have repeated labels then we have to run an initialization algorithm [6,7,8,11]. The protocol (d, h)-PROT works as follows. At the beginning the base station b sends a *start* message to all stations $S\backslash\{b\}$ and sets *connected* := **true** and *level* := 0. In the meantime, remaining stations $S \backslash \{b\}$ are waiting for this message and after receiving it, they set *connected* := **false**. Next, at the phase $j \in \{1, 2, \ldots, h - 1\}$, each non-*connected* station flips a biased *coin* with the probability $p^*(d, h, n, j - 1, \alpha)$ of heads coming up and at the phase $j = h$, each remaining non-*connected* station flips a biased *coin* with the probability 1 of heads coming up. If a non-*connected* station s has just thrown a head ($coin = 1$), then it sends a *search* message at range $r = 2^0, 2^1, \ldots, 2^{\log\lceil\sqrt{d}L\rceil}$, until it receives an echo message from the closest station v that has been connected at level $j - 1$. Therefore, a station s chooses v as its predecessor by fixing the range assignment on $d(s, v)$, i.e. $\mathcal{R}(s) = d(s, v)$. At this point, a station s sets *connected* := true and *level* := j. Once a station s is connected, it will be waiting for messages from non-*connected* stations at level $j + 1$ just like connected stations at level $j - 1$ are waiting for messages from non-*connected* stations at level j. So, if a connected station at level $j - 1$ receives a *search* message from station v at level j then it sends an echo message containing information about the closest to v connected station.

The probability of heads coming up at level $j + 1$, we define as

$$p^*(d, h, j, n, \alpha) = n^{-\lambda(d,h,j,\alpha)}, \text{ where } \lambda(d, h, j, \alpha) = \frac{\sum_{i=1}^{h-j-1}(d/\alpha)^i}{\sum_{i=1}^{h-j}(d/\alpha)^i}. \quad (3)$$

Let assume that a station $s \in S$ throws a head at level $1 \leq j \leq h$. Then, it starts sending messages at increasing distances $r = 2^0, 2^1, \ldots, 2^{\log\lceil\sqrt{d}L\rceil}$ and it stops when it receives an echo message from some station v. In this case, the station s sets its range assignment, i.e. $\mathcal{R}(s) = d(s, v)$. Therefore, its maximal sending range r cannot be larger then $2 \cdot \mathcal{R}(s)$ since at each time we double the range distance r. Moreover, at level $j + 1 \leq h$ the station s will send echo message at ranges equal to those of the corresponding descendant.

We conclude this section by providing the following theorem, which has been proved in [2].

Theorem 3 ([2]). *Let $\hat{\mathcal{R}}(s)$ be the maximal transmission range used by a station $s \in S$ during any level of the protocol and $\mathcal{R}$ be the range assignment returned by the protocol. Then, it holds that*

$$cost(\hat{\mathcal{R}}) = \sum_{s \in S} \hat{\mathcal{R}}(s)^\alpha = \Theta(cost(\mathcal{R})).$$

4   Probabilistic Analysis

In this section, we show that the expected energy cost of the protocol (d, h)-PROT is within $O(1)$ of an optimal solution. Let S be a set of stations. Without loss of generality assume that $S = \{1, \ldots, n\}$, where n is the number of stations, and that a station with identifier n is the base station b. We consider that all stations are uniformly randomly distributed on a d-cube $Q = \{(x_1, x_2, \ldots, x_d) \in \mathbb{R}^d : 0 \leq x_1 \leq L, 0 \leq x_2 \leq L, \ldots, 0 \leq x_d \leq L\}$ of side length L. Let $\{P_s : s \in S\}$ be an independent collection of random variables denoting coordinates of stations in Q. At each level, each station has to decide whether it will be connected at this level or not. Let $\{B_s^j : s \in S \setminus \{b\}, 1 \leq j \leq h\}$ be a collection of independent random variables denoting the result of tossing a biased coin by a station s at level j. We say that a station $s \in S$ has thrown a head if $B_s^j = 1$ and a tail if $B_s^j = 0$. Thus, we obtain that

$$\mathbb{P}[B_s^j = 1] = p^*(d, h, j - 1, n, \alpha), \quad \mathbb{P}[B_s^j = 0] = 1 - p^*(d, h, j - 1, n, \alpha), \quad (4)$$

for $s \in S \setminus \{b\}$, $j = 1, 2, \ldots, h$. Let H_s be a random variable denoting a level at which a station $s \in S \setminus \{b\}$ becomes connected. A station can become connected at level j if it has just tossed a head and before that it had tossed only tails. Therefore, the probability that a station $s \in S \setminus \{b\}$ becomes connected at level j is defined as

$$\mathbb{P}[H_s = j] = \begin{cases} 0, & \text{for } j = 0, \\ \mathbb{P}[B_s^1 = 0, \ldots, B_s^{j-1} = 0, B_s^j = 1], & \text{for } 1 \le j \le h - 1, \\ 1 - \mathbb{P}[B_s^1 = 0, \ldots, B_s^{h-1} = 0], & \text{for } j = h. \end{cases} \quad (5)$$

Lemma 1. *For every station $s \in S \setminus \{b\}$ and $j = 1, 2, \ldots, h - 1$, the following inequalities hold*

$$\mathbb{P}[B_s^1 = 0, \ldots, B_s^{j-1} = 0, B_s^j = 1] \le p^*(d, h, 0, n, \alpha), \quad (6)$$

$$\mathbb{P}[B_s^1 = 0, \ldots, B_s^j = 0] \le 1 - p^*(d, h, 0, n, \alpha), \quad (7)$$

for either $\alpha = 1, d \ge 2$ or $\alpha = 2, d = 2$.

Proof. By the assumption that random variables B_s^j have independent distributions and by (4), we get

$$\mathbb{P}[B_s^0 = 0, \ldots, B_s^{j-1} = 0] = \prod_{k=1}^{j} \mathbb{P}[B_1^k = 0] = \prod_{k=0}^{j-1} (1 - p^*(d, h, k, n, \alpha)),$$

for every station $s \in S$. The probability $p^*(d, h, k, n, 1)$ is given by (3). Therefore, for the case $\alpha = 1$ and $d \ge 2$, we have

$$p^*(d, h, j, n, 1) = n^{-\lambda(d,h,j,1)}, \text{ where } \lambda(d, h, j, 1) = \frac{d^h - d^{j+1}}{d^{h+1} - d^{j+1}}.$$

Now we can show that

$$\lambda(d, h, j - 1, 1) > \lambda(d, h, j, 1), \quad \text{for } j = 1, 2, \ldots, h - 2. \quad (8)$$

Notice that $\lambda(d, h, j, 1) = \frac{1 - d^{j+1}/d^h}{d - d^{j+1}/d^h}$. Let $f(x) = \frac{1-x}{d-x} = 1 - \frac{d-1}{d-x}$. Then $\lambda(d, h, j, 1) = f(\frac{d^{j+1}}{d^h})$, for $j = 1, 2, \ldots, h - 2$. Since $f(x)$ is a strictly decreasing function on the interval $[0, 1]$, thus the inequality (8) holds, so we obtain

$$1 \ge 1 - p^*(d, h, 0, n, 1) \ge 1 - p^*(d, h, 1, n, 1) \ge \ldots \ge 1 - p^*(d, h, h - 2, n, 1). \quad (9)$$

Thus the inequality (7) holds. Similarly, we obtain

$$\mathbb{P}[B_s^0 = 0, \ldots, B_s^{j-1} = 0, B_s^j = 1] = \mathbb{P}[B_s^j = 1] \prod_{k=1}^{j} \mathbb{P}[B_1^k = 0]$$

$$= p^*(d, h, h - 2, n, 1) \prod_{k=0}^{j-1} (1 - p^*(d, h, k, n, 1)).$$

By inequality (9), we finally get

$$\mathbb{P}[B_s^0 = 0, \ldots, B_s^{j-1} = 0, B_s^j = 1] \leq p^*(d, h, 0, n, 1).$$

Thus the inequality (6) holds.

Now let us consider the case $\alpha = 2$ and $d = 2$. We have the probability

$$p^*(2, h, j, n, 2) = n^{-\lambda(2,h,j,2)}, \quad \text{where } \lambda(2, h, j, 2) = \frac{h - j - 1}{h - j}.$$

It follows immediately that

$$\lambda(2, h, j - 1, 2) > \lambda(2, h, j, 2), \quad \text{for} \quad j = 1, 2, \ldots, h - 2.$$

Therefore, similarly we can show that inequalities (6) and (7) holds.   □

Lemma 2. *Let $\mu \in \mathbb{R}^+$ and $d \geq 2$ be a natural number. Then*

$$\int_0^\infty e^{-\mu x^d} dx = \Gamma\left(1 + \frac{1}{d}\right) \cdot \frac{1}{\sqrt[d]{\mu}}, \tag{10}$$

where $\Gamma(x)$ stands for the Gamma function [10].

Proof. By substitution $t = \mu x^d$, we obtain that

$$\int_0^\infty e^{-\mu x^d} dx = \int_0^\infty \frac{1}{d} \mu^{-\frac{1}{d}} t^{\frac{1}{d}-1} e^{-t} dt.$$

Since $\Gamma(x) = \int_0^\infty t^{x-1} e^{-t} dt$ and $\Gamma(x + 1) = x \cdot \Gamma(x)$ the proof is complete.   □

Lemma 3. *The expected cost of the range assignment $\mathcal{R}$ returned by the protocol (d, h)-PROT satisfies the following bound*

$$\mathbb{E}[cost(\mathcal{R})] = \begin{cases} O\left(L \cdot n^{1 - \frac{1}{d} + \frac{d-1}{d^{h+1} - d}}\right), & \text{if } d \geq 2, \alpha = 1, \\ O\left(L^2 \cdot n^{\frac{1}{h}}\right), & \text{if } d = 2, \alpha = 2, \end{cases}$$

for any fixed $h \geq 2$.

Proof. Let h be a fixed number of hops toward the base station b, either greater then or equal 2. Let $\{D_s^j : s \in S \setminus \{b\}, 1 \leq j \leq h\}$ be a collection of random variables denoting a minimal distance between a station s which is connected at level j and any station at level less then j, for $s \in S$ and $j = 1, 2, \ldots, h$. Therefore, by the assumption that the base station b has an identifier n, we obtain

$$\mathbb{E}[cost(\mathcal{R})] = \mathbb{E}\left[\sum_{s \in S \setminus \{b\}} \mathcal{R}(s)^\alpha\right] = \sum_{s \in S \setminus \{b\}} \mathbb{E}[\mathcal{R}(s)^\alpha]$$

$$= \sum_{i=1}^{n-1} \sum_{j=1}^{h} \mathbb{E}[\mathcal{R}(i)^\alpha | H_i = j]\mathbb{P}(H_i = j).$$

Case $\alpha = 1$ and $d \geq 2$. To calculate a conditional expected cost $\mathcal{R}(i)$ for a station $i \in S \setminus \{b\}$, given that station i has been connected at level j, we need to compute the following integral

$$\mathbb{E}[\mathcal{R}(i) | H_i = j] = \int_0^\infty \mathbb{P}[D_i^j > \rho | H_i = j] d\rho.$$

At first we calculate a conditional probability that a station i at level j has a minimal distance to any station at level $j - 1$ greater then ρ. Let $C_{i,\rho} = \{q \in Q : d(P_i, q) < \rho\}$ be the area of the intersection between the disk of radius ρ centered at P_i and the d-cube Q. Then, for $j = 1$, we obtain

$$\mathbb{P}[D_i^1 > \rho | H_i = 1] = \mathbb{P}[P_b \notin C_{i,\rho}] = 1 - \frac{|C_{i,\rho}|}{L^2}.$$

Furthermore, for a station $i \in S \setminus \{b\}$ the minimal distance D_i^j between a station i that is connected at level j ($H_i = j$) and any station at level less then j, is greater then ρ at level j, if a station $s \in S \setminus \{b, i\}$ has not been connected yet, i.e. $B_s^1 = 0, \ldots, B_s^{j-1} = 0$ or its distance from station i is greater then ρ, i.e. $P_s \notin C_{i,\rho}$. Since station i can be directly connected to the base station b, thus the distance from b has to also be greater then ρ, i.e. $P_b \notin C_{i,\rho}$. Thus, for $j = 2, 3, \ldots, h$, we have that $\mathbb{P}[D_i^j > \rho | H_i = j]$ is equal to

$$\mathbb{P}[(\forall s \in S \setminus \{i, b\})((B_s^1 = 0, \ldots, B_s^{j-1} = 0) \vee P_s \notin C_{i,\rho}) \wedge P_b \notin C_{i,\rho}].$$

Let $\beta_j = (1 - \mathbb{P}[B_1^0 = 0, \ldots, B_1^{j-1} = 0])$. Then, by De Morgan's laws and by the assumption that B_i^j and P_s are independent random variables, we obtain

$$\mathbb{P}[D_i^j > \rho | H_i = j] = \prod_{s \in S \setminus \{i, b\}} \mathbb{P}[(B_s^1 = 0, \ldots, B_s^{j-1} = 0) \vee P_s \notin C_{i,\rho}]\mathbb{P}[P_b \notin C_{i,\rho}]$$

$$= \prod_{s \in S \setminus \{i, b\}} \left(1 - (1 - \mathbb{P}[B_s^1 = 0, \ldots, B_s^{j-1} = 0])\,\mathbb{P}[P_s \in C_{i,\rho}]\right)\mathbb{P}[P_b \notin C_{i,\rho}]$$

$$= \left(1 - \frac{|C_{i,\rho}|}{L^2}\right)\left(1 - \beta_j \frac{|C_{i,\rho}|}{L^2}\right)^{n-2} \leq \left(1 - \beta_j \frac{|C_{i,\rho}|}{L^2}\right)^{n-1},$$

for $j = 2, 3, \ldots, h$. Let $(x)_+ = \max\{0, x\}$. Then, by the fact that $|C_{i,\rho}| \geq \rho^d/2^{d+1}$ if $\rho \leq 2\sqrt{d}L$, we obtain the following inequality

$$\mathbb{P}[D_i^j > \rho | H_i = j] \leq \begin{cases} (1 - \frac{\rho^d}{2^{d+1}L^d})_+, & \text{for } j = 1, \\ (1 - \beta_j \frac{\rho^d}{2^{d+1}L^d})_+^{n-1}, & \text{for } 2 \leq j \leq h. \end{cases} \quad (11)$$

Now, we can calculate conditional expected $\mathcal{R}$ for $j = 1$ as follows

$$\mathbb{E}[\mathcal{R}(i)|H_i = j] = \int_0^\infty \left(1 - \frac{\rho^d}{2^{d+1}L^d}\right)_+ d\rho \leq \Gamma\left(1 + \frac{1}{d}\right) 2\sqrt[d]{2}L \quad (12)$$

and for $j = 2, 3, \ldots, h$, we have

$$\mathbb{E}[\mathcal{R}(i)|H_i = j] = \int_0^\infty \left(1 - \beta_j \frac{\rho^d}{2^{d+1}L^d}\right)_+^{n-1} d\rho. \quad (13)$$

By the well-know inequality $1 - x \leq e^{-x}$ and Lemma 2, the integral (13) can be bounded above by

$$\int_0^\infty \left(1 - \beta_j \frac{\rho^d}{2^{d+1}L^d}\right)_+^{n-1} d\rho \leq \int_0^\infty e^{-((n-1)\beta_j\rho^d)/(2^{d+1}L^d)} d\rho$$

$$= \Gamma\left(1 + \frac{1}{d}\right) \frac{2\sqrt[d]{2}L}{\sqrt[d]{\beta_j(n-1)}}$$

and it follows by Lemma 1 that the probability β_j can be lower bounded by

$$\beta_j = 1 - \mathbb{P}[B_1^0 = 0] \cdot \ldots \cdot \mathbb{P}[B_1^{j-1} = 0] \geq 1 - (1 - p^*(d, h, 0, n, 1)) = p^*(d, h, 0, n, 1).$$

Combining these two inequalities leads to that $\mathbb{E}[\mathcal{R}(i)|H_i = j]$ can be bounded above by

$$\mathbb{E}[\mathcal{R}(i)|H_i = j] \leq \frac{\Gamma\left(1 + \frac{1}{d}\right) 2\sqrt[d]{2}L}{\sqrt[d]{\beta_j(n-1)}} \leq \frac{8L}{\sqrt[d]{p^*(d, h, 0, n, 1)(n-1)}}$$

$$\leq \frac{8L}{\sqrt[d]{n^{-(d^h-1-1)/(d^h-1)}}(n-1)} \leq \frac{8L}{\sqrt[d]{\frac{1}{2}n^{1-(d^h-1-1)/(d^h-1)}}}$$

$$\leq \frac{8\sqrt[d]{2}L}{n^{(d^h-2(d-1))/(d^h-1)}}, \quad (14)$$

for $j = 2, 3, \ldots, h$. The first inequality for $\mathbb{E}[\mathcal{R}(i)|H_i = j]$ follows from the fact that $\Gamma(1 + \frac{1}{d}) \leq 2$ if $d \geq 2$. By Lemma 1 the overall expected energy cost can be bounded above by

$$\mathbb{E}[cost(\mathcal{R})] = \sum_{i=1}^{n-1} \sum_{j=1}^{h} \mathbb{E}[\mathcal{R}(i)|H_i = j]\mathbb{P}(H_i = j)$$

$$\leq n^{-(d^h-1-1)/(d^h-1)} \sum_{j=1}^{h} \sum_{i=1}^{n-1} \mathbb{E}[\mathcal{R}(i)|H_i = j].$$

We calculate $\sum_{j=1}^{h} \mathbb{E}[\mathcal{R}(i)|H_i = j]$ by splitting it into two terms as follows

$$\mathbb{E}[\mathcal{R}(i)|H_i = 1] + \sum_{j=2}^{h} \mathbb{E}[\mathcal{R}(i)|H_i = j].$$

Therefore, by inequalities (12) and (14) respectively, we obtain

$$\sum_{j=1}^{h} \mathbb{E}[\mathcal{R}(i)|H_i = j] \le 8 \cdot L + \frac{8\sqrt[d]{2}(h-1)L}{n^{(d^{h-2}(d-1))/(d^h-1)}} \le 8\sqrt[d]{2}(h-1)L,$$

for $i = 1, 2, \ldots, n-1$. Finally, the overall expected energy cost can be bounded above by

$$\mathbb{E}[cost(\mathcal{R})] \le 8\sqrt[d]{2} \cdot h \cdot (n-1) \cdot L \cdot n^{-(d^{h-1}-1)/(d^h-1)}$$
$$= O\left(L \cdot n^{1-(d^{h-1}-1)/(d^h-1)}\right) = O\left(L \cdot n^{d^h(d-1)/(d^{h+1}-d)}\right)$$
$$= O\left(L \cdot n^{1-\frac{1}{d}+\frac{d-1}{d^{h+1}-d}}\right).$$

Case $\alpha = 2$ and $d = 2$. To calculate a conditional expected cost $\mathcal{R}(i)^2$ for a station $i \in S \setminus \{b\}$, given that station i has been connected at level j, we need to compute the following integral

$$\mathbb{E}[\mathcal{R}(i)^2|H_i = j] = 2\int_0^\infty \rho \cdot \mathbb{P}[D_i^j > \rho|H_i = j]d\rho,$$

where $\mathbb{P}[D_i^j > \rho|H_i = j]$ by (11) for $d = 2$ is given by

$$\mathbb{P}[D_i^j > \rho|H_i = j] \le \begin{cases} (1 - \frac{\rho^2}{8L^2}), & \text{for } j = 1, \\ (1 - \beta_j \frac{\rho^2}{8L^2})^{n-1}, & \text{for } 2 \le j \le h. \end{cases}$$

Now we can calculate a conditional expected $\mathcal{R}$ for $j = 1$. Thus, we obtain

$$\mathbb{E}[\mathcal{R}(i)^2|H_i = j] = 2\int_0^\infty \rho \cdot \left(1 - \frac{\rho^2}{8L^2}\right)_+ d\rho \le 8L^2 \qquad (15)$$

and for $j = 2, 3, \ldots, h$, we have

$$\mathbb{E}[\mathcal{R}(i)^2|H_i = j] = 2\int_0^\infty \rho \cdot \left(1 - \beta_j \frac{\rho^2}{8L^2}\right)_+^{n-1} d\rho. \qquad (16)$$

By the inequality $1 - x \le e^{-x}$ and since $\int_0^\infty xe^{-\mu x^2} dx = \frac{1}{2\mu}$, for any $\mu \in \mathbb{R}_+$, we obtain

$$2\int_0^\infty \rho \cdot \left(1 - \beta_j \frac{\rho^2}{8L^2}\right)_+^{n-1} d\rho \le 2\int_0^\infty \rho \cdot e^{-((n-1)\beta_j \rho^2)/(8L^2)} d\rho$$
$$= \frac{8L^2}{\beta_j(n-1)}.$$

By Lemma 1 for $\alpha = 2, d = 2$, we obtain the following inequality

$$\beta_j = 1 - \mathbb{P}[B_1^0 = 0] \cdot \ldots \cdot \mathbb{P}[B_1^{j-1} = 0] \geq 1 - (1 - p^*(d, h, 0, n, 2)) = p^*(d, h, 0, n, 2)$$

and thus $\mathbb{E}[\mathcal{R}(i)^2 | H_i = j]$ can be bounded above by

$$\mathbb{E}[\mathcal{R}(i)^2 | H_i = j] \leq \frac{8L^2}{\beta_j(n-1)} \leq \frac{8L^2}{p^*(d, h, 0, n, 2)(n-1)}$$

$$\leq \frac{8L^2}{n^{-(h-1)/h}(n-1)} \leq \frac{8L^2}{\frac{1}{2}n^{1-(h-1)/h}} \leq \frac{16L^2}{n^{1/h}}. \qquad (17)$$

Thus, by Lemma 1 for $\alpha = 2, d = 2$ we have

$$\mathbb{E}[cost(\mathcal{R})] = \sum_{i=1}^{n-1} \sum_{j=1}^{h} \mathbb{E}[\mathcal{R}(i)^2 | H_i = j] \mathbb{P}(H_i = j)$$

$$\leq n^{-(h-1)/h} \sum_{j=1}^{h} \sum_{i=1}^{n-1} \mathbb{E}[\mathcal{R}(i)^2 | H_i = j].$$

Therefore, by inequalities (15) and (17) respectively, we obtain

$$\sum_{j=1}^{h} \mathbb{E}[\mathcal{R}(i)^2 | H_i = j] \leq 8 \cdot L^2 + \frac{16(h-1)L^2}{n^{1/h}} \leq 32(h-1)L^2,$$

for $i = 1, \ldots, n-1$. Finally, we get

$$\mathbb{E}[cost(\mathcal{R})] \leq 32 \cdot h \cdot (n-1) \cdot L^2 \cdot n^{-(h-1)/h}$$

$$= O\left(L^2 \cdot n^{1-(h-1)/h}\right) = O\left(L^2 \cdot n^{\frac{1}{h}}\right).$$

$\square$

We are ready to prove the claimed approximation ratio of the protocol (d, h)-PROT.

Theorem 4. *The protocol (d, h)-PROT on random instances, for either $\alpha = 1, d \geq 2$ or $\alpha = 2, d = 2$ and any fixed $h \geq 2$ achieves an approximation ratio of $O(1)$ in expectation.*

Proof. By Theorem 3, we have that the energy cost of the protocol (d, h)-PROT is $\Theta(cost(\mathcal{R}))$. Then, combining Theorem 1 and Lemma 3 concludes the proof.

5 Conclusion

In this paper we presented the protocol (d, h)-PROT for d-DIM h-HOPS CONVERGECAST problem. The (d, h)-PROT is an extension of the existing protocol h-PROT proposed in [2] on a d-dimensional space. The main result of our work is the formal proof that the protocol (d, h)-PROT achieves the approximation ratio of $O(1)$ in expectation for any $d, h \geq 2$ on random instances. Therefore, in particular our result answers positively the open problem stated by Clementi *et al.* in [2].

References

1. Alfandari, L., Paschos, V.T.: Approximating minimum spanning tree of depth 2. International Transactions in Operational Research 6(6), 607–622 (1999)
2. Clementi, A.E.F., Ianni, M.D., Lauria, M., Monti, A., Rossi, G., Silvestri, R.: Protocol for the Bounded-Hops Converge-Cast in Ad-Hoc Networks. In: Kunz, T., Ravi, S.S. (eds.) ADHOC-NOW 2006. LNCS, vol. 4104, pp. 60–72. Springer, Heidelberg (2006)
3. Clementi, A.E.F., Ianni, M.D., Monti, A., Rossi, G., Silvestri, R.: Divide and Conquer is almost optimal for the Bounded-Hop Accumulation in Ad Hoc Wireless Networks. In: The Proceedings of SIROCCO (2005)
4. Clementi, A.E.F., Ianni, M.D., Monti, A., Rossi, G., Silvestri, R.: Experimental Analysis of Practically Efficient Algorithms for Bounded-Hop Accumulation in Ad-Hoc Wireless Networks. In: IPDPS 2005. Proceedings of the 19th IEEE International Parallel and Distributed Processing Symposium (2005)
5. Clementi, A.E.F., Penna, M.D., Silvestri, R.: On the Power Assignment Problem in Radio Networks. Mobile Networks and Applications (MONET) 9, 125–140 (2004)
6. Cichoń, J., Kutyłowski, M., Zawada, M.: Adaptive Initialization Algorithm for Ad Hoc Radio Networks with Carrier Sensing. In: Nikoletseas, S.E., Rolim, J.D.P. (eds.) ALGOSENSORS 2006. LNCS, vol. 4240, pp. 35–46. Springer, Heidelberg (2006)
7. Cichoń, J., Kutyłowski, M., Zawada, M.: Initialization for Ad Hoc Radio Networks with Carrier Sensing and Collision Detection. In: Kunz, T., Ravi, S.S. (eds.) ADHOC-NOW 2006. LNCS, vol. 4104, pp. 308–320. Springer, Heidelberg (2006)
8. Cai, Z., Lu, M., Wang, X.: Distributed Initialization Algorithm for Single-Hop Ad Hoc Networks with Minislotted Carrier Sensing. IEEE Transactions Parallel and Distributed Systems 14(5), 516–528 (2003)
9. Gouveia, L.: Using the Miller-Tucker-Zemlin constraints to formulate a minimal spanning tree problem with hop constraints. Computer and Operations Research 22, 959–970 (1995)
10. Gradshteyn, I.S., Ryzhik, I.M., Jeffrey, A.: Table of Integrals, Series, and Products, 6th edn. Academic Press, London (2000)
11. Nakano, K., Olariu, S.: Randomized Initialization Protocols for Ad Hoc Networks. IEEE Transactions Parallel and Distributed Systems 11(7), 749–759 (2000)

Correlation, Coding, and Cooperation in Wireless Sensor Networks

Samar Agnihotri, Pavan Nuggehalli, and H.S. Jamadagni

CEDT, Indian Institute of Science, Bangalore - 560012, India
{samar,pavan,hsjam}@cedt.iisc.ernet.in

Abstract. We consider a single-hop data-gathering sensor network, consisting of a set of sensor nodes that transmit data periodically to a base-station. We are interested in maximizing the lifetime of this network. With our definition of network lifetime and the assumption that the radio transmission energy consumption forms the most significant portion of the total energy consumption at a sensor node, we attempt to enhance the network lifetime by reducing the transmission energy budget of sensor nodes by exploiting three system-level opportunities.

We pose the problem of maximizing lifetime as a max-min optimization problem subject to the constraint of successful data collection and limited energy supply at each node. This turns out to be an extremely difficult optimization to solve. To reduce the complexity of this problem, we allow the sensor nodes and the base-station to interactively communicate with each other and employ *instantaneous decoding* at the base-station. The chief contribution of the paper is to show that the computational complexity of our problem is determined by the complex interplay of various system-level opportunities and challenges.

Keywords: Sensor networks, Lifetime maximization, Multi-access networks, Joint source-channel coding, Data correlation, Slepian-Wolf coding, Scheduling.

1 Introduction

Many extant and future applications of the sensor networks demand long operational lifetimes of the networks. The sensor nodes constituting these networks are supposed to be tiny devices with modest energy reserves and limited capabilities of computation and communication. In such situations, the key challenge is to devise communication protocols and network architectures that are conscious of these constraints, yet provide long operational lifetimes for such networks. A broad concensus exists on devising aggressive system-level strategies impacting many layers of the protocol stack to meet the lifetime requirement of extant and future sensor networks.

We consider a single-hop, data-gathering wireless sensor network. Sensor nodes periodically sample a field and transmit the data directly to a base-station. We define network lifetime as the time until the first node in the network runs out of the energy to communicate. While somewhat pessimistic, this definition is well-suited for the applications where the failure of even a single node can have disastrous consequences on the network's performance (for example, reducing coverage or causing network partitioning.) This definition also has the benefit of being simple and popular [1, 2]. It should be noted that our model of sensor networks is also applicable to the networks, where

M. Kutyłowski et al. (Eds.): ALGOSENSORS 2007, LNCS 4837, pp. 83–98, 2008.

the nodes are organized as clusters [3] and the network lifetime maximization problem reduces to multiple cluster lifetime maximization problems.

Our definition of network lifetime implies that to maximize it, we need to minimize the energy consumption at sensor nodes. Sensor nodes expend energy in sensing, computing, and communication. We neglect the energy consumed by the nodes in sensing and computing as sensing costs are independent of the communication strategy employed and computing costs are often negligible compared to communication costs.

The energy expended by a sensor node in communication has two components: reception energy and transmission energy. The energy consumed in reception depends on the number of bits received and the per bit energy cost to keep the receiver circuitry energized. The transmission energy depends on a number of factors such as number of bits to transmit, transmit power levels, receiver sensitivity, channel state (including path loss due to distance and fading), and the channel coding scheme employed. We assume that optimal channel coding is employed. We ignore the energy cost of reception, as it can be easily incorporated in our proposed model. This reduces the problem of maximizing the network lifetime to the problem of minimizing the energy cost of transmission of the nodes.

We propose to exploit three system-level opportunities to minimize the radio transmit energy cost of sensor nodes. First, sensor data in a data-gathering network is spatially and temporally correlated. In [4], Slepian and Wolf show that it is possible to compress a set of correlated sources down to their joint entropy, without explicit communication among sources. Recent advances in distributed source coding [5,6], allow us to take advantage of data correlation to reduce the number of transmitted bits, with concomitant savings in energy. Second, [7] shows that channel coding can be used to reduce transmission energy by increasing the transmission time. Finally, sensor nodes are supposed to be cooperative. This collaborative nature allow us to exploit the first two opportunities to minimize transmission energy consumption at nodes.

We pose the problem of maximizing network lifetime as an optimization problem, subject to the constraint of successful data collection and limited energy supply at each node. However, to our surprise, we find that even with ample simplification, analyzing the performance of our model of the sensor network is far from easy. The chief contribution of the paper is to illustrate the dependence of the computational complexity of our problem on the complex interplay of above system-level opportunities. We provide various insights, analyses, and algorithms for several scenarios. In some situations, our problem admits a greedy solution while in others, the problem is shown to be $\mathcal{NP}$-hard.

There is much related work in this area. Energy conscious networking strategies have been proposed mainly at the MAC [8] and routing layer [1, 9, 10, 11, 12, 13]. Our study was motivated by the work in [14, 15], which explicitly incorporate aggregation costs in gathering sensor data. However, both [14] and [15] are interested in minimizing total energy expenditure, as opposed to maximizing network lifetime. There the optimal solution is shown to be a greedy solution based on ordering sensors according to their distance (which reflects data aggregation cost) from the base-station. However, we show that this solution is not optimal for maximizing network lifetime. This paper generalizes the work in [16], placing it in a wider context and lending it a firm theoretical basis.

2   System Model

We consider a network of N battery operated sensor nodes strewn uniformly in a coverage area and communicating directly with the base-station. The network operates in a time-division multiple access (TDMA) mode. In each time-slot, every sensor is allotted a certain amount of time to communicate its data to the base-station. We only consider the spatial correlation among sensor readings in a time-slot and ignore the temporal correlation over different time-slots, as latter can be easily incorporated in our work for data sources satisfying the Asymptotic Equipartition Property (AEP).

Initially, sensor node k, $1 \leq k \leq N$, has access to E_k units of energy. The wireless channel between sensor k and the base-station is described by a path loss factor d_k, which captures various channel effects such as distance induced attenuation, shadowing, and multipath fading. For simplicity, we assume d_k's to be constant. This is reasonable for static networks and also in the scenarios where the path loss parameter varies slowly and so, can be accurately tracked.

The general sensor network lifetime maximization problem is to solve joint source-channel coding problem for multi-access networks. We assume the separation between source and channel coding, albeit at some loss of optimality. This suboptimality occurs as it is well-known that, in general, the source-channel separation does not hold for the multi-access joint source-channel coding problem [17]. So, in general, Slepian-Wolf coding followed by optimal channel coding is not optimal for this problem. Also, turning a multiple-access channel into an array of orthogonal channels by using a suitable MAC protocol (TDMA in our case) is well-known to be a suboptimal strategy, as the set of rates achievable with orthogonal access is strictly contained in the Ahlswede-Liao capacity region [18]. However, despite these sub-optimalities, we argue like [19] and [20] that there are strong economic gains in the deployment of networks based on such technologies, due to the low complexity and cost of existing solutions, as well as the availability of vast experience in the design and operation of such systems.

The problem addressed in this paper is to find optimal rate (the number of bits to transmit) and transmission time vectors for all nodes, which maximize network lifetime. Both the rate and time allocation are constrained. The rate allocation should fall within the Slepian-Wolf achievable rate region and the sum of transmission times should be less than the period of a time-slot (which is taken to be unity.) However, finding the optimal rates and time allocations are computationally challenging problems as the Slepian-Wolf achievable rate region for N nodes is defined by $2^N - 1$ constraints and within a time-slot, the nodes can cooperate by varying their transmission times.

As a pragmatic step forward, we propose to solve the optimal rate allocation problem by allowing the sensor nodes and the base-station to communicate interactively and employing *instantaneous decoding* at the base-station. Allowing interactive communication, lets us model the "sensor nodes - base-station" communication as the "multiple correlated informants - single recipient" communication, discussed in [6]. "Instantaneous decoding" implies that the decoding at the base-station be instantaneous in the sense that once a particular node has been polled, the data generated at that node is recovered at the base-station before the next node is polled. The introduction of these two concepts, reduces the network lifetime maximization problem to finding an optimal scheduling strategy (optimal polling order and transmission time allocation.)

We assume that the communication takes place over N binary, error-free channels, where each channel connects a sensor with the base-station. A sensor node and the base-station can communicate back and forth over the channel connecting them, but the sensors cannot communicate directly with each other [6]. We concern ourselves only with the problem of maximizing the average lifetime of the network, as in [6]-IV.B and make following assumptions, which without altering the general nature of our results, simplify the presentation of our ideas. The sensor nodes and the base-station communicate using the two message protocol proposed in [6]-II.E. We use entropies, instead of average code lengths. Also, we ignore the energy costs of reception at the sensors and both, the transmission and reception energy costs at the base-station.

Notation. Let Π be the set of all permutations of the set, $\{1, 2, \ldots, N\}$. Let $\pi(k)$ denote the kth polled node in the schedule $\pi \in \Pi$. Instantaneous decoding implies that the number of information bits transmitted per slot by $\pi(k)$ is the conditional entropy of the data source at $\pi(k)$, given the data generated by all previously polled nodes, it is denoted as $h_{\pi(k)}$. Let $t_{\pi(k)}$ be the corresponding transmission time alloted to node k.

3 General Channel Scenario

In this section, we consider the sensor network lifetime maximization problem, exploiting all three system-level opportunities discussed in "Introduction". The spatial correlation in sensor data helps in reducing the number of bits that a node has to transmit, improving the network lifetime. The cooperative nature of the sensor nodes and channel coding can then be exploited to improve the network lifetime further by varying the transmission times of the nodes. For example, highly correlated nodes can finish their transmissions sooner, allowing weakly correlated nodes more time to transmit in order to improve system lifetime. However, in such a scenario, we not only have to find the optimal scheduling order, but also the optimum transmission time for each node.

We consider, both, the static and dynamic schedules. In static scheduling, a fixed sensor polling order is followed in all time-slots until the network dies. Under dynamic scheduling, possibly different schedules are employed in different time-slots to poll the sensor nodes. More specifically, we employ *offline* dynamic scheduling, where before the actual data-gathering starts, the base-station computes the optimum set of schedules and the number of slots for which each schedule is used; rather than *online* dynamic scheduling, where at the beginning of every time-slot, the base-station computes the optimum sensor polling schedule, based on its latest estimate of network state.

We consider the channel coding scenarios where the transmission energy is the convex decreasing function of the transmission time [7]. Let $f(h, x)d$ be the energy required to transmit h bits of information in x units of time with path loss factor d. We model the energy function $f(h, x)$ as follows.

1. $f(h, x)$ is a strictly *decreasing* continuous positive convex function in x.
2. $\lim_{x \to 0} f(h, x) = \infty$

For the rest of this section, we assume that $f(h, x)$ is:

$$f(h, x) = x(2^{\frac{h}{x}} - 1) \tag{1}$$

3.1 Static Scheduling

In static scheduling, each permutation, $\pi \in \Pi$ corresponds to a TDMA schedule. Let L_π be the network lifetime achievable under schedule π. Note that lifetime is integer-valued, but we treat it as a real number. The optimal static schedule is the solution to the following optimization problem.

$$\max_{\pi \in \Pi} L_\pi (= \min_{1 \leq k \leq N} \frac{E_k}{f(h_{\pi(k)}, t_{\pi(k)})d_k}) \tag{2}$$

$$t_{\pi(k)} \geq 0, 1 \leq k \leq N,$$

$$\sum_{k=1}^{N} t_{\pi(k)} = 1.$$

However, using the "Channel Aware" algorithm proposed in [16], for every schedule π, we can compute the maximum lifetime and the corresponding transmission times allocation vector $\{t_{\pi(k)}\}_{k=1}^{N}$. Also, for this transmission time allocation, all sensor nodes achieve the same lifetime. So, (2) reduces to:

$$\max_{\pi \in \Pi} L_\pi \tag{3}$$

$$\text{s.t. } f(h_{\pi(k)}, t_{\pi(k)})d_k = \frac{E_k}{L_\pi}, 1 \leq k \leq N$$

$$t_{\pi(k)} \geq 0, 1 \leq k \leq N$$

$$\sum_{k=1}^{N} t_{\pi(k)} = 1.$$

For a given energy consumption function, the computational complexity of the problem in (2) or (3) depends largely on the sensor data correlation model. The following three examples amply illustrate this.

For our first two examples, let us consider the first model of spatial correlation in sensor data introduced in [21], with $\alpha_1 = 1.0, \beta_1 = 1.0$. So, let us define $B(X_i/X_j)$, the number of bits that the node i has to send when the node j has already sent its bits to the base-station, as follows:

$$B(X_i/X_j) = \begin{cases} \lceil d_{ij} \rceil \text{ if } d_{ij} \leq n \\ n \text{ if } d_{ij} > n, \end{cases} \tag{4}$$

where X_i be the random variable representing the sampled sensor reading at node $i \in \{1, \ldots, N\}$, n is the maximum number of bits that a node can send, and d_{ij} denotes the distance between nodes i and j.

Let us define $B(X_i/X_1, \ldots, X_{i-1})$, the conditional information when more than one node has already sent its information to the base-station, as follows:

$$B(X_i/X_1, \ldots, X_{i-1}) = \min_{1 \leq j < i} B(X_i/X_j) \tag{5}$$

Example 1. Assume that $E_k/d_k = constant$, for every sensor $k, 1 \leq k \leq N$. This assumption models the scenario, where given the equal energies of all the nodes, the distance of the base-station from any node is much more than any inter-node distance. This assumption is also valid when a roving base-station gathers the sensor data. For the energy consumption model of (1), the time to transmit depends only on the number of bits that a node sends to the base-station. So, a sensor polling schedule that minimizes $B(X_1, \ldots, X_N)$, the total number of bits sent by N nodes, also minimizes the sum of transmission times of all the nodes, and subsequently maximizes the network lifetime.

Theorem 1. *A greedy scheme provides the optimal solution.*

Proof. Starting with any node as the first node of the schedule, choose the next node in the schedule to be the node that minimizes the conditional number of bits. However, for the correlation model of (4) and (5), this amounts to finding the nearest node. So, the schedule that selects the nearest neighbor as the next node to be polled is the optimum schedule and we call it "Nearest Neighbor Next (NNN)" schedule. For a desired value of network lifetime, the NNN schedule will give the smallest value of $B(X_1, X_2, \ldots, X_N)$ and the smallest sum of the transmission times, so using the "Channel Aware" algorithm proposed in [16], we can prove that this schedule is optimum.   □

Example 2. Consider the general problem, where for sensor $k, 1 \leq k \leq N$, we do not assume that $E_k/d_k = constant$.

Theorem 2. *The static scheduling problem in (3) is $\mathcal{NP}$-hard.*

Proof. An arbitrary instance of "Shortest Hamiltonian Path" problem can be reduced to this problem as in the proof in Example 3.   □

Example 3. Let us consider a spatial correlation model, where the sensor data is modeled by Gaussian random field [15]. Thus, we assume that an N dimensional *jointly Gaussian multivariate distribution* $f(\mathbf{X})$ models the spatial data $\mathbf{X}$ of N sensor nodes.

$$f(\mathbf{X}) = \frac{1}{(\sqrt{2\pi})^N \det(K)^{1/2}} e^{-\frac{1}{2}(\mathbf{X}-\mu)^T K^{-1}(\mathbf{X}-\mu)}, \tag{6}$$

where K is the (positive definite) covariance matrix of $\mathbf{X}$, and μ the mean vector. The diagonal entries of K are the variances $K_{ii} = \sigma_i^2$. The rest of K_{ij} depend on the distance between the nodes i and j: $K_{ij} = \sigma^2 \exp(-a d_{i,j}^2)$.

Let us assume that the data at each sensor node is quantized with the same quantization step, then differential entropy differs from entropy by a constant and without any loss of generality, can be used instead of entropy.

Theorem 3. *The static scheduling problem in (3) is $\mathcal{NP}$-hard.*

Proof. Consider the decision version of this problem: does there exist a schedule π, for which the network achieves the lifetime L, with the following constraints?

$$f(h_{\pi(k)}, t_{\pi(k)})d_k = \frac{E_k}{L}, 1 \leq k \leq N \tag{7}$$

$$t_{\pi(k)} \geq 0, 1 \leq k \leq N$$

$$\sum_{k=1}^{N} t_{\pi(k)} \leq 1.$$

Next, we reduce an arbitrary instance of "Shortest Hamiltonian Path (SHP) Problem" [22] over Euclidean and complete graph to some instance of the problem in (7). The rest of the proof is given in the Appendix. □

3.2 Dynamic Scheduling

In this section, we explore if the network lifetime can be increased by employing possibly different schedules in different time-slots. Two or more schedules can collaborate by having the nodes use non-optimal transmit energies over two or more time-slots to increase the lifetime of the network.

We have a total of $N!$ schedules. If only m, $1 \leq m \leq N!$ schedules cooperate, then there are $C(N!, m)$ sets of m schedules. For a given set of m schedules, let τ_{π_i} denote the number of time-slots for which schedule $\pi_i, i \in \{1, \ldots, m\}$ is employed. Then, the optimal network lifetime L is the solution to the following problem.

$$L = \max_{\substack{m \\ 1 \leq m \leq N!}} \max_{\substack{[\pi_1, \ldots, \pi_m] \\ \pi_1, \ldots, \pi_m \in \Pi}} \sum_{i=1}^{m} \tau_{\pi_i} \tag{8}$$

$$\text{s. t. } \sum_{i=1}^{m} f(h_{\pi_i(k)}, t_{\pi_i(k)}) d_k \tau_{\pi_i} \leq E_k, 1 \leq k \leq N$$

$$\sum_{k=1}^{N} t_{\pi_i(k)} = 1, 1 \leq i \leq m$$

Specifically for $m = N!$, we have:

$$L = \max \sum_{i=1}^{N!} \tau_{\pi_i} \tag{9}$$

$$\text{s. t. } \sum_{i=1}^{N!} f(h_{\pi_i(k)}, t_{\pi_i(k)}) d_k \tau_{\pi_i} \leq E_k, 1 \leq k \leq N$$

$$\sum_{k=1}^{N} t_{\pi_i(k)} = 1, 1 \leq i \leq N!$$

Note that for $m = 1$, (8) reduces to the static scheduling problem in (3). So, the computational complexity of this problem cannot be any less than that of the static scheduling problem, which is proven to be $\mathcal{NP}$-hard for the most of scenarios of interest. However, we still wonder if the dynamic scheduling can improve the network lifetime. In the following, we prove that even for the simplest case of the network of two nodes, it is indeed so, and then generalize this results for $N > 2$ or $m > 2$.

Theorem 4. *For $N = 2$, dynamic scheduling performs better than the optimal static scheduling.*

Proof. For the network of two nodes, let us consider two schedules: π_1, where node 1 is polled before node 2, and π_2, where the nodes are polled otherwise. Using the "Channel Aware" algorithm in [16], for a given polling schedule, we can find the optimal allocation of the transmission times such that both the nodes spend same amount of energy and die simultaneously. Assume that for schedule π_1, this happens when the node 1 transmits for t units of time and node 2 transmits for $1 - t$ units of time. Similarly, for schedule π_2, let the corresponding times be t' and $1 - t'$. Let h and $h_{1|2}$ denote the entropies of first and second node polled in the schedule, respectively. So, for schedule π_1: $h_1 = h, h_2 = h_{2|1} = h_{1|2}$ and for schedule π_2: $h_1 = h_{1|2}, h_2 = h$. Given the optimality of t and t' for the schedules π_1 and π_2 respectively, we have:

$$\text{For schedule } \pi_1 : f(h,t)d_1 = f(h_{1|2}, 1 - t)d_2, \tag{10}$$

$$\text{For schedule } \pi_2 : f(h_{1|2}, t')d_1 = f(h, 1 - t')d_2. \tag{11}$$

Assuming that the schedule π_1 is the optimum static schedule, we have:

$$f(h,t)d_1 = f(h_{1|2}, 1 - t)d_2 \leq f(h_{1|2}, t')d_1 = f(h, 1 - t')d_2$$

Let us consider a 2D Cartesian plot, where the energy consumption E_1 of the node 1 and E_2 of the node 2, define the X and Y axes, respectively. In this plot, we draw the energy consumption curves for the schedules π_1 and π_2, for different values of $t, 0 < t < 1$ and $t', 0 < t' < 1$ respectively. Given the form of these curves, it is easy to verify that these curves are convex and intersect at only one point.

The equation of the "equal energy line" that passes through the pair of points $(f(h,t)$ $d_1, f(h_{1|2}, 1 - t)d_2)$ and $(f(h_{1|2}, t')d_1, f(h, 1 - t')d_2)$, is:

$$E_1 = E_2 \tag{12}$$

Now, let us also consider a line passing through the points $(f(h,r)d_1, f(h_{1|2}, 1 - r)d_2)$ and $(f(h_{1|2}, s)d_1, f(h, 1 - s)d_2)$ on the curves corresponding to the schedules π_1 and π_2, respectively with $0 < r < 1$ and $0 < s < 1$. The equation for such a line is:

$$E_2 = f(h_{1|2}, 1 - r)d_2 + m(r,s)(E_1 - f(h,r)d_1 \tag{13}$$

$$\text{where } m(r,s) = \frac{f(h, 1 - s)d_2 - f(h_{1|2}, 1 - r)d_2}{f(h_{1|2}, s)d_1 - f(h,r)d_1}.$$

At the point of intersection of these two lines, we have:

$$E_1 = E_2 \tag{14}$$

$$= \frac{f(h_{1|2}, 1 - r)d_2 - m(r,s)f(h,r)d_1}{1 - m(r,s)}$$

$$= \frac{f(h_{1|2}, 1 - r)f(h_{1|2}, s)d_2 - f(h, 1 - s)f(h,r)d_2}{f(h_{1|2}, s) - f(h,r)}$$

To prove that the dynamic scheduling can perform better than the static scheduling, we must prove that there exists at least one pair of values (r, s), such that:

$$E_1(r,s) < f(h,t)d_1 \tag{15}$$

$$E_2(r,s) < f(h_{1|2}, 1 - t)d_2$$

Substituting the expressions of $E_1(r, s)$ and $E_2(r, s)$ from (14) in (15), and using the properties of the energy consumption function, we prove that for all (r, s) such that

$$r < t \tag{16}$$

$$\frac{ht - (h - h_{1|2})}{h_{1|2}} < s,$$

the dynamic scheduling outperforms static scheduling.   □

This result implies that two schedule can cooperate to give longer network lifetime compared to optimum static schedule. Now, we generalize this result.

Theorem 5. $L_m \geq L_{m-1}$, where $L_m, 2 \leq m \leq N!$, denotes the optimum network lifetime for m schedule cooperation.

Proof. Omitted for brevity.   □

Theorem 6. *For every* N, *there exists an optimum set* S^* *of* m_N^* *schedules, such that for any other optimum set* S *of* $m, m_N^* < m \leq N!$ *schedules,* $L_m = L_{m_N^*}$.

Proof. Omitted for brevity.   □

These two theorems together imply that as m varies from 1 to $N!$, there exists a unique set of m optimal schedules, which maximizes the network lifetime for the m schedule cooperation. Also, for $1 < m \leq m_N^*$, $L_m \geq L_{m-1}$, but when $m_N^* < m \leq N!$, $L_m = L_{m_N^*}$.

4   Small Rate Region Approximation

Let us assume that transmission rate is linearly proportional to signal power. This assumption is motivated by Shannon's AWGN capacity formula, which is approximately linear for low data rates. The low rate assumption implies, as shown below, that transmit energy is independent of transmission time. Hence, the optimal time allocation problem is trivial and we only need to find the optimal polling schedule.

In the following, we show that our assumption admits greedy, polynomial time solutions for the optimal rate allocation problem, in contrast to the generally $\mathcal{NP}$-hard solutions of the last section. However, this reduction in the computational complexity is achieved by settling for possibly lower network lifetimes, as the nodes can no more cooperate to improve the network lifetime by varying their transmission times, as in the previous section.

We assume that for the small data rates, the energy consumption function for node $\pi(k)$ is $f(h_{\pi(k)}, t_{\pi(k)})d_k = h_{\pi(k)}d_k$. For example, by inverting Shannon's AWGN channel capacity formula, we get [7]:

$$f(h_{\pi(k)}, t_{\pi(k)})d_k = t_{\pi(k)}(2^{\frac{h_{\pi(k)}}{t_{\pi(k)}}} - 1)d_k.$$

So, for the small data rates, this gives for some constant $c \geq 0$:

$$f(h_{\pi(k)}, t_{\pi(k)})d_k \approx t_{\pi(k)}(1 + c\frac{h_{\pi(k)}}{t_{\pi(k)}} - 1)d_k$$

$$= ch_{\pi(k)}d_k.$$

4.1   Static Scheduling

Under the "small rate region approximation", the static scheduling problem in (2) reduces to:

$$\max_{\pi \in \Pi} \min_{1 \leq k \leq N} \frac{E_{\pi(k)}}{h_{\pi(k)}d_{\pi(k)}} \tag{17}$$

The objective function represents the lifetime of node $\pi(k)$. In the following, we describe a polling strategy, which we call "Minimum Cost Next (MCN)". When a scheduling decision needs to be taken, MCN chooses that node (among the unpolled ones), which consumes the smallest fraction of its initial energy. The MCN schedule is denoted by π^{MCN}.

Algorithm. MCN

---

1   S : set of all N nodes.
2   A : set of nodes whose polling order has been computed.
3   Initialization: $A = \phi$, $k = 1$.
4   **while** $(k \leq N)$
5       $\pi^{MCN}(k) = \arg\min_{i \in S - A} \frac{d_i h(X_i | A)}{E_i}$.
6       $A = A \cup \pi^{MCN}(k)$.
7       $k = k + 1$.

Theorem 7. *MCN schedule is the optimal static schedule.*

Proof. We describe an iterative procedure to modify a given schedule into MCN schedule such that network lifetime does not decrease. This proves that MCN schedule is optimal, since we can start with an optimal schedule and iteratively apply our procedure to obtain MCN schedule, at no loss of network lifetime. Let π^{OLD} be any schedule that differs from π^{MCN} first in the mth position, that is:

$$\pi^{OLD}(k) = \pi^{MCN}(k), \quad 1 \leq k \leq m - 1 \tag{18}$$
$$\pi^{OLD}(m) \neq \pi^{MCN}(m).$$

Then, there exists a number l such that $\pi^{OLD}(l) = \pi^{MCN}(m)$, $l > m$. We construct a new schedule π^{NEW} by modifying π^{OLD} as follows:

$$\pi^{NEW}(k) = \pi^{MCN}(k), \quad 1 \leq k \leq m \tag{19}$$
$$\pi^{NEW}(k) = \pi^{OLD}(k - 1), \quad m < k \leq l$$
$$\pi^{NEW}(k) = \pi^{OLD}(k), \quad l < k \leq N$$

In words, in π^{NEW}, we poll π^{MCN} for first m-slots, followed by π^{OLD} for next $N-m$ slots. For any schedule π, let $g(\pi, k)$ be the lifetime of the kth polled node. Let us define:

$$g(\pi, k) = \frac{E_{\pi(k)}}{h_{\pi(k)} d_{\pi(k)}}.$$

To establish that π^{NEW} is at least as good as π^{OLD}, we need to show that

$$\min_{1 \le k \le N} g(\pi^{NEW}, k) \ge \min_{1 \le k \le N} g(\pi^{OLD}, k) \tag{20}$$

From construction, it follows that for $1 \le k \le m-1$ and $l+1 \le k \le N$

$$g(\pi^{NEW}, k) = g(\pi^{OLD}, k)$$

So, it suffices to show that

$$\min_{m \le k \le l} g(\pi^{NEW}, k) \ge \min_{m \le k \le l} g(\pi^{OLD}, k) \tag{21}$$

Using the fact that conditioning reduces entropy, we have

$$\min_{m+1 \le k \le l} g(\pi^{NEW}, k) \ge \min_{m+1 \le k \le l} g(\pi^{OLD}, k) \tag{22}$$

Moreover, the MCN construction ensures that

$$g(\pi^{NEW}, m) \ge g(\pi^{OLD}, m) \tag{23}$$

Equations (22) and (23), together imply (21), proving the theorem.   □

Now we state without proof, two properties of the MCN schedule. First, the MCN schedule not only maximizes the minimum lifetime, but also maximizes all lifetimes from 2nd minimum lifetime to N^{th} minimum lifetime. This is desirable in the situations, where the network has to continue to operate even when one or more nodes die out. Second, MCN solution is Pareto-optimal, as given an MCN schedule, no other schedule can help increase any node's lifetime without decreasing some other node's lifetime.

The MCN algorithm is a greedy algorithm and its worst-case computational complexity is $\mathcal{O}(N^2)$.

4.2 Dynamic Scheduling

In this section, we explore how network lifetime can be increased by employing multiple schedules under the "small rate region approximation". Under this assumption, as the general static scheduling problem in (2) reduced to (17), the general dynamic scheduling problem in (9) reduces to:

$$L = \max \sum_{\pi \in \Pi} \tau_\pi \tag{24}$$

$$\text{s.t. } \sum_{\pi \in \Pi} h_{\pi(k)} d_k \tau_\pi \le E_k, \ 1 \le k \le N$$

τ_π is the number of slots for which the schedule π is used. Once more, the constraints ensure that the time assignment is feasible for each node with respect to its energy capability. Also, as in subsection 3.2, (24) can be treated as a linear program. A dynamic schedule, τ, is given by the set $\{\tau_\pi\}$.

Given $N!$ variables, in general, there seems to be no easy way to solve (24). However, the special nature of our problem can be exploited to yield efficient methods. It is reasonable to assume that sensor readings at a node are strongly correlated only with neighboring nodes. For any schedule, the energy consumed by a node will then depend primarily on its relative order in the schedule with respect to its neighbors. This *clustering* phenomena leads to considerable reduction in computational effort. Moreover, the max-min nature of our optimization problem simplifies the search for an optimal schedule and allows us to easily determine when a schedule is optimal. These properties are exploited in "Lifetime Optimal Clustering ALgorithm (LOCAL)", introduced in [16].

Algorithm. LOCAL

---

1    $k = 0,\ C_j^0 = \pi(j),\ R_k = N.$
2   **repeat**
3   **for** $(j = 1$ **to** $R_k)$
4   Solve eqn (24) for nodes in C_j^k.
5   Find lifetime $L_j^k = \min_{i \in C_j^k} l_i^k$.
6            $s = \arg\min_{1 \le j \le R_k} L_j^k.$
7   **if** $(s == R_k)$
8   **then** $L = L_s^k$. BREAK.
9   **else**
10   Merge C_s^k and C_{s+1}^k.
11                    $R_k = R_k - 1.$
12                    $k = k + 1.$

Theorem 8. *LOCAL is optimal.*

Proof. Suppose LOCAL terminates in stage k. Then:

$$L_{R_k}^k = \min_{1 \le j \le R_k} L_j^k$$

We prove that LOCAL is optimal by showing that, under any dynamic schedule, the minimum of the lifetimes of nodes in cluster, $C_{R_k}^k$, cannot exceed $L_{R_k}^k$. Since network lifetime can only be less than or equal to cluster lifetime, optimality is proved. Under LOCAL, all nodes which do not belong to cluster $C_{R_k}^k$ are scheduled before $C_{R_k}^k$. Hence nodes in cluster $C_{R_k}^k$ are *maximally* conditioned. This fact, along with the cluster level optimization performed in stage k, ensures that no dynamic schedule can assure a greater lifetime for nodes in cluster $C_{R_k}^k$.   □

In the worst case, when all the nodes merge into one single cluster, LOCAL reduces to (24) and this determines the worst-case complexity of LOCAL.

5   Conclusions

We have considered the problem of maximizing the lifetime of a data-gathering wireless sensor network. Our contribution differs from previous research in two respects. Firstly, we proposed a practical source-channel coding framework to mitigate the energy cost of radio transmission. Secondly, we have explicitly maximized network lifetime. To the best of our knowledge, both these aspects have not been explored in the context of sensor networks previously.

The chief contribution of our work lies in explicitly demonstrating the dependence of the computational complexity of the sensor network lifetime maximization problem, exploiting a few system-level opportunities, on the relationship between energy consumption and transmission rate as well as model assumptions about correlation in sensor data, path loss, and initial energy reserves.

References

1. Chang, J., Tassiulas, L.: Energy conserving routing in wireless ad-hoc networks. In: Proc. IEEE INFOCOM 2000, Tel-Aviv, Israel (March 2000)
2. Kang, I., Poovendran, R.: Maximizing static network lifetime of wireless broadcast adhoc networks. In: Proc. IEEE ICC 2003, Anchorage, AK (May 2003)
3. Heinzelman, W.R., Chandrakasan, A., Balakrishnan, H.: Energy-efficient communication protocol for wireless microsensor networks. In: Proc. HICSS 2000, Maui, HI (2000)
4. Slepian, D., Wolf, J.K.: Noiseless coding of correlated information sources. IEEE Trans. Inform. Theory IT-19(4), 471–480 (1973)
5. Xiong, Z., Liveris, A.D., Cheng, S.: Distributed source coding for sensor networks. IEEE Signal Processing Mag. 21 (September 2004)
6. Agnihotri, S., Nuggehalli, P., Rao, R.: Enhancing sensor network lifetime using interactive communication. In: Proc. IEEE ISIT 2007, Nice, France (June 2007)
7. Uysal-Biyikoglu, E., Prabhakar, B., El Gamal, A.: Energy-efficient packet transmission over a wireless link. IEEE Trans. Networking 10(4), 487–499 (2002)
8. Yei, W., Heidemann, J., Estrin, D.: An energy-efficient MAC protocol for wireless sensor networks. In: Proc. 21st Intl. Ann. Joint Conf. of the IEEE Computer and Communication Societies (2002)
9. Bhardwaj, M., Chandrakasan, A.P.: Bounding the lifetime of sensor networks via optimal role assignments. In: Proc. IEEE INFOCOM 2002, New York City (June 2002)
10. Singh, S., Woo, M., Raghavendra, C.S.: Power-aware routing in mobile ad hoc networks. In: Proc. ACM MOBICOM 1998, Dallas, TX (October 1998)
11. Rodoplu, V., Meng, T.: Minimum energy mobile wireless networks. IEEE JSAC 17(8), 1333–1344 (1999)
12. Sadagopan, N., Krishnamachari, B.: Maximizing data extraction in energy-limited sensor networks. In: Proc. IEEE INFOCOM 2004, Hong Kong (March 2004)
13. Li, Q., Aslam, J., Rus, D.: Online power-aware routing in wireless ad-hoc networks. In: Proc. ACM MOBICOM 2001, Rome, Italy (June 2001)
14. Baek, S.J., de Veciana, G., Su, X.: Minimizing energy consumption in large-scale sensor networks through distributed data compression and hierarchical aggregation. IEEE JSAC 22(6), 1130–1140 (2004)
15. Cristescu, R., Lozano, B.B., Vetterli, M.: On network correlated data gathering. In: Proc. IEEE INFOCOM 2004, Hong Kong (March 2004)

16. Agnihotri, S., Nuggehalli, P., Jamadagni, H.S.: On maximizing lifetime of a sensor cluster. In: Proc. WoWMoM 2005, Taormina, Italy (June 2005)
17. Cover, T.M., El Gamal, A., Salehi, M.: Multiple access channels with arbitrarily correlated sources. IEEE Trans. Inform. Theory IT-26(6), 648–657 (1980)
18. Cover, T.M., Thomas, J.: Elements of Information Theory. John Wiley & Sons, New York (1991)
19. Barros, J., Servetto, S.D.: Network information flow with correlated sources. IEEE Trans. Inform. Theory IT-52(1), 155–170 (2006)
20. Kawadia, V., Kumar, P.R.: A cautionary perspective on cross-layer design. IEEE Wireless Comm. Mag. 12(1), 3–11 (2005)
21. Agnihotri, S.: New models for the correlation in sensor data. pre-print available at arXiv: cs.IT/0702035
22. Rubin, F.: A search procedure for Hamilton paths and circuits. Journal of ACM 21(4), 576–580 (1974)

Appendix. Proof of Theorem 3

Proof. For the given instance of SHP problem, interpret the edge cost between nodes i and j, as the spatial distance between the nodes i and j of our problem. So, as we visit a node k in the SHP tour π, we can compute the conditional entropy $h_{\pi(k)}$ of that node using the knowledge of the model for spatial correlation among the sensor nodes as well as the history of the tour so far. Using the first constraint of (7), we can compute the minimum time $t_{\pi(k)}$ that node $\pi(k)$ needs to transmit $h_{\pi(k)}$ bits of information to the base-station. So, for every schedule π, we can compute the sum of the minimum transmission times.

Let us consider an Euclidean, complete graph of $N = 5$ nodes with vertex set $[A, B, C, D, E]$ and symmetric edge costs, as in figure 1. We have chosen this graph only to illustrate the main idea of the reduction, but our approach can also be used for bigger, yet *similar* networks. Let us consider two schedules $ABCDE$ and $ABDCE$.

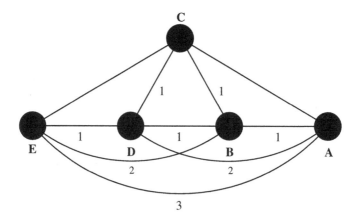

Fig. 1. Network graph for Example 2. Note that $d_{AC} = d_{CE} = \sqrt{3}$

Let us assumes that the length of Hamiltonian path d_{ABCDE} for schedule $ABCDE$ is less than the Hamiltonian path length d_{ABDCE} for the schedule $ABDCE$. In the following, we prove that with this assumption, the sum of transmission times for the schedule $ABCDE$ is less than the corresponding sum for the schedule $ABDCE$. For the spatial correlation model of interest (6), for every schedule, we can compute the conditional entropy of every node based on all the nodes visited previously [18]. For example, if the schedule is $ABCDE$ and X_A, X_B, X_C, X_D denote data samples of the nodes $A, B, C,$ and D, respectively, then:

$$h(X_A) = \frac{1}{2} \log(2\pi e \sigma_A^2),$$

$$h(X_B/X_A) = \frac{1}{2} \log((2\pi e) \det(K_{AB})),$$

$$h(X_C/X_A, X_B) = \frac{1}{2} \log\left((2\pi e)\frac{\det(K_{ABC})}{\det(K_{AB})}\right),$$

$$h(X_D/X_A, X_B, X_C) = \frac{1}{2} \log\left((2\pi e)\frac{\det(K_{ABCD})}{\det(K_{ABC})}\right),$$

where $K_{AB}, K_{ABC},$ and K_{ABCD}, respectively denote the covariance matrices of vectors $(X_A, X_B), (X_A, X_B, X_C),$ and (X_A, X_B, X_C, X_D).

Let us define a few quantities as follows:

$$\Sigma_{[A,B,C]} = e^{-2\alpha d_{AB}^2} + e^{-2\alpha d_{AC}^2} + e^{-2\alpha d_{BC}^2},$$

$$\Sigma_{[A,B,D]} = e^{-2\alpha d_{AB}^2} + e^{-2\alpha d_{AD}^2} + e^{-2\alpha d_{BD}^2},$$

$$\Sigma_{[A,B,C,D]} = \Sigma_{[A,B,C]} + e^{-2\alpha d_{AD}^2} + e^{-2\alpha d_{BD}^2} + e^{-2\alpha d_{CD}^2}.$$

Let $\sigma_i = 1, i \in [A, B, C, D, E]$. Then, expanding the determinants above gives:

$$h(X_A) = \frac{1}{2} \log(2\pi e), \tag{25}$$

$$h(X_B/X_A) = 1 + \frac{1}{2} \log(1 - \sigma^2 e^{-2\alpha d_{AB}^2}) - h(X_A), \tag{26}$$

$$h(X_C/X_A, X_B) = \frac{1}{2} + \frac{1}{2} \log \frac{1 - \sigma^2 \Sigma_{[A,B,C]} + 2\sigma^3 e^{-\alpha(d_{AB}^2 + d_{AC}^2 + d_{BC}^2)}}{1 - \sigma^2 e^{-2\alpha d_{AB}^2}},$$

$$\approx \frac{1}{2} + \frac{1}{2} \log \frac{1 - \sigma^2 \Sigma_{[A,B,C]}}{1 - \sigma^2 e^{-2\alpha d_{AB}^2}}, \tag{27}$$

$$h(X_D/X_A, X_B, X_C) \approx \frac{1}{2} + \frac{1}{2} \log \frac{1 - \sigma^2 \Sigma_{[A,B,C,D]}}{1 - \sigma^2 \Sigma_{[A,B,C]}}. \tag{28}$$

Let us denote the transmission times of the nodes $A, B, C, D,$ and E under schedule $ABCDE$ as $t_A, t_B, t_C, t_D,$ and t_E, respectively. For the schedule $ABDCE$, let the corresponding times be $t_A', t_B', t_C', t_D',$ and t_E', respectively. Note that $t_A = t_A', t_B = t_B', t_E = t_E'$. Now,

$$t_A + t_B + t_C + t_D + t_E < t_A' + t_B' + t_C' + t_D' + t_E' \tag{29}$$

$$\text{if } t_C + t_D < t_C' + t_D'. \tag{30}$$

Assume that the transmission time $t_{\pi(k)}$ of nodes $k \in [A, B, C, D, E]$, is exponentially dependent on the corresponding entropy $h_{\pi(k)}$ of the node[1]. Substituting the values of $t_C, t_D, t'_C,$ and t'_D in (30) and then a little algebraic manipulation of the resulting expression, gives:

$$
\left[\frac{1 - \sigma^2 \Sigma_{[A,B,C]}}{1 - \sigma^2 e^{-2\alpha d^2_{AB}}}\right]^{\frac{1}{2}} + \left[\frac{1 - \sigma^2 \Sigma_{[A,B,C,D]}}{1 - \sigma^2 \Sigma_{[A,B,C]}}\right]^{\frac{1}{2}}
$$
$$
<
$$
$$
\left[\frac{1 - \sigma^2 \Sigma_{[A,B,D]}}{1 - \sigma^2 e^{-2\alpha d^2_{AB}}}\right]^{\frac{1}{2}} + \left[\frac{1 - \sigma^2 \Sigma_{[A,B,C,D]}}{1 - \sigma^2 \Sigma_{[A,B,D]}}\right]^{\frac{1}{2}}
$$
(31)

Consider the inequality:

$$
\left(\frac{B}{A}\right)^{\frac{1}{2}} + \left(\frac{C}{B}\right)^{\frac{1}{2}} < \left(\frac{B'}{A}\right)^{\frac{1}{2}} + \left(\frac{C}{B'}\right)^{\frac{1}{2}},
$$
(32)

which always holds if $0 < C < B < B' < A < 1$. So, (31) will hold true if we can prove that

$$
1 - \sigma^2 \Sigma_{[A,B,C]} < 1 - \sigma^2 \Sigma_{[A,B,D]}
$$
$$
\text{Or } \sigma^2 \Sigma_{[A,B,C]} > \sigma^2 \Sigma_{[A,B,D]}
$$
$$
\text{Or } e^{-2\alpha d^2_{AC}} + e^{-2\alpha d^2_{BC}} > e^{-2\alpha d^2_{AD}} + e^{-2\alpha d^2_{BD}}
$$
$$
\text{Or } e^{-2\alpha(d^2_{AC} + d^2_{BC})} > e^{-2\alpha(d^2_{AD} + d^2_{BD})}
$$
$$
\text{Or } d^2_{AC} + d^2_{BC} < d^2_{AD} + d^2_{BD}
$$
(33)

For the graph in figure 1, it is obvious that (33) holds. Also for this graph, it is straightforward to show that $d_{ABCDE} < d_{ABDCE} \implies d^2_{AC} + d^2_{BC} < d^2_{AD} + d^2_{BD}$.

So, if a schedule has smaller Hamiltonian path length, then the corresponding sum of the transmission times will be smaller too. This implies that the solution of SHP gives the smallest value of the sum of the transmission times. So, for the schedule that gives shortest Hamiltonian path, we can compute the sum of the transmission times and if this sum is less than 1, then we have at least one schedule that gives the lifetime L.   □

---

[1] This follows from numerically solving the first constraint in (7) for t_i, assuming $\frac{E_k}{d_k} = const, 1 \le k \le N$.

Local Approximation Algorithms for Scheduling Problems in Sensor Networks

Patrik Floréen, Petteri Kaski, Topi Musto, and Jukka Suomela

Helsinki Institute for Information Technology HIIT
Department of Computer Science, University of Helsinki
P.O. Box 68, FI-00014 University of Helsinki, Finland
firstname.lastname@cs.helsinki.fi

Abstract. We study fractional scheduling problems in sensor networks, in particular, sleep scheduling (generalisation of fractional domatic partition) and activity scheduling (generalisation of fractional graph colouring). The problems are hard to solve in general even in a centralised setting; however, we show that there are practically relevant families of graphs where these problems admit a *local* distributed approximation algorithm; in a local algorithm each node utilises information from its constant-size neighbourhood only. Our algorithm does not need the spatial coordinates of the nodes; it suffices that a subset of nodes is designated as *markers* during network deployment. Our algorithm can be applied in any marked graph satisfying certain bounds on the marker density; if the bounds are met, guaranteed near-optimal solutions can be found in constant time, space and communication per node. We also show that auxiliary information is necessary—no local algorithm can achieve a satisfactory approximation guarantee on unmarked graphs.

1 Introduction

The scalability of distributed algorithms presents a basic hurdle to the envisaged large-scale implementations of sensor networking, in particular due to the bounded resources of the individual network nodes. Simply put, if we want to operate arbitrarily large sensor networks, we cannot apply network control algorithms where the communication or computation per node increases with increasing network size. Indeed, if each individual network node is powered by a battery with bounded capacity, there is always a threshold size for the network beyond which the energy consumption for network control exceeds the battery capacity of a node.

1.1 Local Algorithms

In this work we study *local algorithms* [1], where each node must operate solely based on information that was available at system startup within a constant-size neighbourhood of the node. A local algorithm provides an extreme form of scalability: assuming constant-size input per node, the communication, space

M. Kutyłowski et al. (Eds.): ALGOSENSORS 2007, LNCS 4837, pp. 99–113, 2008.

and time complexity of a local algorithm is constant per node. Thus, a local algorithm scales to an arbitrarily large (or even infinite) resource-constrained network. We detail the model of computation in Sect. 3.

A local algorithm is clearly the ideal choice for sensor networks, but even from a theoretical perspective it is not immediate whether such algorithms can exist for practical computational problems arising in network control. This work shows that various NP-hard *scheduling problems* admit *deterministic, local approximation algorithms* provided that the network meets certain assumptions on its structure.

1.2   Scheduling Problems

We study two basic scheduling problems pertinent to sensor networks: *sleep scheduling*, a fractional packing problem, and *(conflict-free) activity scheduling*, a fractional covering problem. Both problems can be formulated as a linear program (LP), but the number of variables in the LP can be exponential in the size of the network; both problems are NP-hard to solve even in a centralised setting.

To ease the exposition, we present the scheduling problems first in a centralised setting; the requirements for a proper distributed solution are detailed together with the local computational model in Sect. 3. We require a few preliminaries to present the definitions. All graphs are undirected. We model the network topology by a *communication graph* $\mathcal{G} = (V_{\mathcal{G}}, E_{\mathcal{G}})$, where each node $v \in V_{\mathcal{G}}$ corresponds to a sensor device and each edge $\{u, v\} \in E_{\mathcal{G}}$ indicates that u and v can directly communicate with each other. We denote by $d_{\mathcal{G}}(u, v)$ the shortest-path distance (hop count) between nodes $u, v \in V_{\mathcal{G}}$ in $\mathcal{G}$ and extend the notation to subsets $U \subseteq V_{\mathcal{G}}$ by $d_{\mathcal{G}}(U, v) = \min_{u \in U} d_{\mathcal{G}}(u, v)$. For $v \in V_{\mathcal{G}}$ and $r \geq 0$, we define the closed ball of radius r centred at v in $\mathcal{G}$ by $B_{\mathcal{G}}(v, r) = \{u \in V_{\mathcal{G}} : d_{\mathcal{G}}(u, v) \leq r\}$.

Problem 1 (SLEEP SCHEDULING). The input to the problem consists of (i) the communication graph $\mathcal{G}$; (ii) a subgraph $\mathcal{R}$ of $\mathcal{G}$ called the *redundancy graph*; and (iii) a battery capacity $b(v) \geq 0$ for each node $v \in V_{\mathcal{R}}$. Each edge $\{u, v\} \in E_{\mathcal{R}}$ indicates that the nodes u and v are pairwise redundant; each node may sleep only if at least one of its neighbours in $\mathcal{R}$ is awake. The valid sets of awake nodes are precisely the dominating sets of $\mathcal{R}$. For a dominating set D, we define $D(v) = 1$ if $v \in D$ and $D(v) = 0$ if $v \notin D$. Denoting by $x(D)$ the length of the time period associated with the dominating set D, the task in the problem is to maximise the total length $\sum_D x(D)$ subject to $\sum_D D(v)x(D) \leq b(v)$ and $x(D) \geq 0$, where v ranges over $V_{\mathcal{R}}$ and D ranges over all the dominating sets of $\mathcal{R}$. To simplify subsequent analysis, we assume that the values $b(v)$ are chosen from a fixed, finite set of nonnegative rational numbers (say, the capacities of the standard batteries in stock); in particular, a constant number of bits per node suffice to encode the input, as it is enough to identify which battery is installed in the device instead of encoding an arbitrary battery capacity.

The sleep scheduling problem is a generalisation of fractional domatic partition. It captures the problem of maximising the lifetime of a battery-powered sensor network by letting each node sleep occasionally, subject to coverage constraints under a pairwise redundancy model [2, 3, 4, 5, 6].

Problem 2 (ACTIVITY SCHEDULING). The input to the problem consists of (i) the communication graph $\mathcal{G}$; (ii) a subgraph $\mathcal{C}$ of $\mathcal{G}$ called the *conflict graph*; and (iii) an activity requirement $a(v) \geq 0$ for each node $v \in V_\mathcal{C}$. Each edge $\{u, v\} \in E_\mathcal{C}$ indicates that the nodes u and v are mutually conflicting; at most one of the two nodes may be active at any given time. The valid sets of active nodes are precisely the independent sets of $\mathcal{C}$. For an independent set I, we define $I(v) = 1$ if $v \in I$ and $I(v) = 0$ if $v \notin I$. Denoting by $x(I)$ the length of the time period associated with the independent set I, the task in the problem is to minimise the total length $\sum_I x(I)$ subject to $\sum_I I(v)x(I) \geq a(v)$ and $x(I) \geq 0$, where v ranges over $V_\mathcal{C}$ and I ranges over all the independent sets in $\mathcal{C}$. Again, we assume that the values $a(v)$ are chosen from a fixed, finite set of nonnegative rational numbers.

The activity scheduling problem is a generalisation of fractional graph colouring. It captures the problem of minimising the total duration of radio transmissions subject to pairwise interference constraints [7].

1.3   Assumptions on Network Structure

Unfortunately, both scheduling problems just presented are hard to solve exactly or approximately [8, 9, 10], even in a centralised setting. To arrive at problem instances that can be solved approximately in a distributed manner, one must impose constraints on the structure of the communication graph $\mathcal{G}$. Furthermore, to obtain a local approximation algorithm, there is a need to break symmetry between the nodes to obtain any satisfactory approximation guarantee, as we will make apparent in Lemma 1.

An embedding of $\mathcal{G}$ in a low-dimensional ambient space could be used as a remedy for both aforementioned difficulties. Indeed, graphs with geometric constraints (for example, unit-disk graphs) in many cases admit efficient approximation algorithms at least in the centralised case, and the spatial coordinates of the nodes break symmetry. However, equipping the nodes with self-positioning capabilities (such as GPS) may not be feasible in large-scale installations, and neither is it practical to inform each node about its physical location during network deployment.

Rather than rely on a geometric embedding, in this work we investigate a minimalistic solution to break symmetry—one *marker* bit of information per node. Furthermore, we use purely combinatorial constraints on the marked $\mathcal{G}$ to arrive at a locally tractable setting. We characterise the admissible distributions of the nodes with the marker bit set (the *markers*) in $\mathcal{G}$ by nonnegative integer parameters ℓ_1, ℓ_μ, and μ, where $\ell_1 < \ell_\mu$.

Definition 1. *A* $(\Delta, \ell_1, \ell_\mu, \mu)$-*marked graph is a pair* $(\mathcal{G}, M)$, *where* $\mathcal{G}$ *is a graph and* $M \subseteq V_\mathcal{G}$ *is a set of* markers *such that, for all* $v \in V_\mathcal{G}$, *(i) the degree of* v *in* $\mathcal{G}$ *is at most* Δ; *(ii)* $d_\mathcal{G}(M, v) \leq \ell_1$; *and (iii)* $|M \cap B_\mathcal{G}(v, \ell_\mu)| \leq \mu$.

In other words, every node has at most Δ neighbours, there is at least one marker within ℓ_1 hops from any node, and there are at most μ markers within ℓ_μ hops from any node. Examples of marked graphs appear in Sect. 5.

1.4   Contributions

As the main technical contribution, we prove the following theorems in Sect. 4. In both theorems, the marking constraint applies to the communication graph $\mathcal{G}$ only.

Theorem 1. *There is a local* $(1+\epsilon)$-*approximation algorithm for sleep scheduling in* $(\Delta, \ell_1, \ell_\mu, \mu)$-*marked graphs for any* $\epsilon > 4\Delta/\lfloor(\ell_\mu - \ell_1)/\mu\rfloor$.

Theorem 2. *There is a local* $1/(1 - \epsilon)$-*approximation algorithm for activity scheduling in* $(\Delta, \ell_1, \ell_\mu, \mu)$-*marked graphs for any* $\epsilon > 4/\lfloor(\ell_\mu - \ell_1)/\mu\rfloor$.

To contrast these positive results, we also demonstrate that the algorithms in Theorems 1 and 2 make near-optimal use of the marking information. In particular, we present a family of marked graphs where our algorithm for sleep scheduling (respectively, activity scheduling) achieves the approximation ratio $1 + 9\epsilon$ (respectively, $1/(1 - 9\epsilon)$) while no local approximation algorithm can achieve the approximation ratio $1 + \epsilon$ (respectively, $1/(1 - \epsilon)$).

2   Earlier Work

2.1   Local Algorithms

Previous work on local algorithms mainly focuses on combinatorial problems such as independent set and graph colouring. Linial [11] shows that any distributed algorithm requires $\Omega(\log^* n)$ communication rounds to find a maximal independent set or a 3-colouring of a ring with n nodes, implying in particular that no local algorithm exists for these tasks. Naor and Stockmeyer [1] present positive results for *locally checkable labelling* problems; for example, it is possible to 2-colour the nodes of a graph using a local algorithm so that each node has at least one neighbour with a different colour, provided that all nodes have odd degree.

Closer to the present work is the work of Kuhn et al. [12], who present local approximation algorithms for fractional covering and packing problems. However, in their work the size of the LP is polynomial in the size of the network, while the size of the LPs that arise from sleep scheduling and activity scheduling can be exponential.

2.2   Shifting Strategy

The present work can be seen as an extension of a classical design paradigm for geometric approximation algorithms—the *shifting strategy* [13]—into a local, coordinate-free, and nongeometric setting. In a typical application of the shifting strategy [13, 14, 15, 16, 17], one uses a grid to partition the (low-dimensional) geometric space into small cells. Each cell defines a subproblem; for example, the subgraph induced by the nodes which are located within or near the cell. Each subproblem is solved optimally, and the solutions are combined to form a feasible global solution. A number of possible locations for the grid are evaluated and the best one is chosen as the solution.

Unfortunately, there are two basic obstacles hindering the application of the shifting strategy in large-scale distributed systems. First, it has been argued that the shifting strategy is "inherently central" [18]; in particular, the final step involves determining which of the candidate solutions is the best one. Second, a straightforward application of the shifting strategy requires that we know how the input is embedded in an ambient space.

Our previous work [3] partially overcomes the aforementioned obstacles in a specific problem: sleep scheduling. To avoid the need for centralised control, we note that the scheduling problem is of fractional nature: one can take two valid schedules and interleave them in order to obtain another valid schedule. To avoid a global coordinate system, we place markers in the underlying communication graph; the constraints for the locations of the markers are geometric, but the algorithm does not use the locations. The present work generalises this previous work in the following aspects: (i) The algorithm is extended to fractional covering problems in addition to fractional packing problems. (ii) No geometric constraints are required; in particular, $\mathcal{G}$ need not have an embedding in a low-dimensional space. (iii) There is no lower bound for the distance between a pair of markers.

3   Preliminaries

3.1   Model of Computation

We assume a communication graph $\mathcal{G}$ where each node has degree bounded by a constant Δ. Each node in $\mathcal{G}$ executes the same distributed deterministic algorithm.

An algorithm is *local* if there exist a constant L ("the local horizon") such that for every problem instance, each node $v \in V_{\mathcal{G}}$ makes its decisions based on information in the nodes $B_{\mathcal{G}}(v, L)$ only. In the sleep scheduling problem, this information consists of the identifiers of the nodes $B_{\mathcal{G}}(v, L)$, the subset of markers $M \cap B_{\mathcal{G}}(v, L)$, the subgraph of the communication graph $\mathcal{G}$ induced by $B_{\mathcal{G}}(v, L)$, the subgraph of the redundancy graph $\mathcal{R}$ induced by $B_{\mathcal{G}}(v, L) \cap V_{\mathcal{R}}$, and the battery capacity $b(u)$ for each node $u \in B_{\mathcal{G}}(v, L) \cap V_{\mathcal{R}}$. The definition is analogous for activity scheduling.

We assume that the node identifiers form an ordered set. An algorithm cannot access the absolute value of an identifier, but only the ordering of the identifiers. In particular, the identifiers need only be unique in $B_{\mathcal{G}}(v, L)$ for each $v \in V_{\mathcal{G}}$. Therefore our computational model is slightly weaker in comparison with the model used by Linial [11]. (To motivate this weakening, see Naor and Stockmeyer [1, Theorem 3.3].)

With these definitions, the number of bits communicated, stored and processed by any node during the execution of a local algorithm is bounded by a constant. Thus also the time complexity is constant per node.

In the scheduling problems, a node does not report any output; instead, a node executes the schedule it has locally computed by controlling its sleeping (respectively, activity). To enable execution of the schedule, it is assumed that (i) each node has access to a clock and (ii) the clocks are (locally) synchronised.

A local $(1 + \epsilon)$-approximation algorithm for sleep scheduling guarantees that the nodes that are awake form a dominating set of the redundancy graph at any point in time during the first $q/(1 + \epsilon)$ time units, where q is the length of an optimal solution. A local $(1+\epsilon)$-approximation algorithm for activity scheduling guarantees that the nodes that are active form an independent set of the conflict graph at any point in time and each node completes its activity within $(1 + \epsilon)q$ time units, where q is the length of an optimal solution.

3.2   Limitations

The chosen local model of computation is very restrictive. For example, Linial [11] shows that (with respect to a strictly stronger model of computation) no local algorithm can properly 3-colour rings. Thus, it is not surprising that scheduling problems in rings are not approximable by local algorithms.

Lemma 1. *No local algorithm on an unmarked graph has an approximation ratio better than 3 for the sleep scheduling problem or any finite approximation ratio for the activity scheduling problem.*

Proof. Consider an arbitrary local algorithm with local horizon $L \in \mathbb{N}$. Let the communication graph $\mathcal{G}$ be a ring of $6L$ nodes, that is, $V_{\mathcal{G}} = \{0, 1, \ldots, 6L-1\}$ and $E_{\mathcal{G}} = \{\{0, 1\}, \{1, 2\}, \ldots, \{6L - 2, 6L - 1\}, \{6L - 1, 0\}\}$. The node identifiers are ordered by $0 < 1 < \ldots < 6L - 1$. For sleep scheduling, let $\mathcal{R} = \mathcal{G}$ and $b(v) = 1$ for each $v \in V_{\mathcal{R}}$; for activity scheduling, let $\mathcal{C} = \mathcal{G}$ and $a(v) = 1$ for each $v \in V_{\mathcal{C}}$. Now the local neighbourhood $B_{\mathcal{G}}(v, L)$ has the same structure for each node in $U = \{L, L+1, \ldots, 5L - 1\}$. At any point in time, all these nodes have to make the same decision.

In the case of sleep scheduling we can obtain a schedule of length 3 by choosing the congruence classes modulo 3 as the dominating sets and by assigning 1 time unit to each. However, if each node in U makes the same decision in the local algorithm, then all of them have to be awake at any point in time; otherwise, e.g., the node $L + 1$ would not be dominated. Thus if the local algorithm produces a feasible sleep schedule, the length of the schedule is at most 1, implying that the local algorithm cannot guarantee an approximation ratio better than 3.

In the case of activity scheduling we can obtain a schedule of length 2 by choosing the congruence classes modulo 2 as the independent sets and by assigning 1 time unit to each. However, if each node in U makes the same decision in the local algorithm, then none of them can be active at any point in time; otherwise a conflicting pair of nodes $\{L, L+1\}$ would be active simultaneously. Thus, the nodes in U can never complete their activities, implying that the local algorithm cannot guarantee any finite approximation ratio.   □

Therefore one has to incorporate auxiliary information to the communication graph to obtain satisfactory approximation guarantees for scheduling.

4   Local Approximability of Scheduling

We assume that the marked communication graph $(\mathcal{G}, M)$ is a $(\Delta, \ell_1, \ell_\mu, \mu)$-marked graph with $k = \lfloor (\ell_\mu - \ell_1)/\mu \rfloor > 0$. Intuitively, a large k is desirable for a good approximation and a small ℓ_μ is desirable in limiting the computational effort.

4.1   Finding Cells

Each node $v \in V_{\mathcal{G}}$ applies the following algorithm:

FIND-CELLS
1   $d \leftarrow d_{\mathcal{G}}(M, v)$
2   **for** $i \leftarrow 0$ **to** $k\mu - 1$
3   **do** $m(v, i) \leftarrow \min(M \cap B_{\mathcal{G}}(v, d+i))$

First, the node finds the distance d to its nearest marker; note that $d \leq \ell_1$. Then, for each *configuration* $i = 0, 1, \ldots, k\mu - 1$, the node finds the smallest marker within the distance $d + i$; here we use the total order on the identifiers.

We define the *cell* of the marker m in configuration i by $C(m, i) = \{v \in V_{\mathcal{G}} : m(v, i) = m\}$. We say that a node $v \in V_{\mathcal{G}}$ is a *boundary node* in configuration i if v has a neighbour u in $\mathcal{G}$ such that $m(v, i) \neq m(u, i)$. The following lemma captures a key property of the configurations (cf. Floréen et al. [3, Lemma 4]).

Lemma 2. *For any $v \in V_{\mathcal{G}}$, there are at most 4μ configurations i such that v is a boundary node in i.*

To prove Lemma 2, we start with two technical lemmata.

Lemma 3. *For any node $v \in V_{\mathcal{G}}$, there are at most μ different values of $m(v, i)$.*

Proof. On line 3 in FIND-CELLS, it holds that $d + i < \ell_1 + k\mu \leq \ell_\mu$, which implies $m(v, i) \in M \cap B_{\mathcal{G}}(v, \ell_\mu)$ for each configuration i. By definition, $|M \cap B_{\mathcal{G}}(v, \ell_\mu)| \leq \mu$.   □

Lemma 4. *For any node $v \in V_{\mathcal{G}}$, there are at most $\mu - 1$ configurations i such that $m(v, i) \neq m(v, i + 1)$.*

Proof. Consider an arbitrary $v \in V_{\mathcal{G}}$. By Lemma 3, it suffices to show that each distinct value of $m(v, i)$ corresponds to a single interval of configurations i; once $m(v, i)$ changes its value from m_1 to $m_2 \neq m_1$, it never changes back to m_1.

Assume that $m(v, i_1) = m(v, i_2) = m$ for arbitrary m and $i_1 \leq i_2$. Then m is a member of the ball $B_{\mathcal{G}}(v, d_{\mathcal{G}}(M, v) + i_1)$, and m is the smallest marker in the larger ball $B_{\mathcal{G}}(v, d_{\mathcal{G}}(M, v) + i_2)$. Thus, for any $i_1 \leq i \leq i_2$, it holds that m is the smallest smallest marker in $B_{\mathcal{G}}(v, d_{\mathcal{G}}(M, v) + i)$, implying $m(v, i) = m$ for all $i_1 \leq i \leq i_2$.   □

Proof of Lemma 2. Consider an arbitrary $v \in V_{\mathcal{G}}$. By Lemma 4, we can divide the list of configurations $(0, 1, \ldots, k\mu - 1)$ into at most μ intervals, such that $m(v, i)$ is constant within each interval. We now prove that v can be a boundary node at most 4 times on each interval.

This clearly holds for intervals of length at most 4. Next, consider an interval from i_1 to i_2 with $i_2 \geq i_1 + 4$ such that $m(v, i) = m$ for each configuration i with $i_1 \leq i \leq i_2$.

Let $u \in V_{\mathcal{G}}$ be any neighbour of v. Because $d_{\mathcal{G}}(u, v) = 1$, it holds that $|d_{\mathcal{G}}(M, v) - d_{\mathcal{G}}(M, u)| \leq 1$. By construction, $m = m(v, i_1)$ is a marker in $B_{\mathcal{G}}(v, d_{\mathcal{G}}(M, v) + i_1) \subseteq B_{\mathcal{G}}(u, d_{\mathcal{G}}(M, v) + i_1 + 1) \subseteq B_{\mathcal{G}}(u, d_{\mathcal{G}}(M, u) + i_1 + 2)$, and $m = m(v, i_2)$ is the smallest marker in $B_{\mathcal{G}}(v, d_{\mathcal{G}}(M, v) + i_2) \supseteq B_{\mathcal{G}}(u, d_{\mathcal{G}}(M, v) + i_2 - 1) \supseteq B_{\mathcal{G}}(u, d_{\mathcal{G}}(M, u) + i_2 - 2)$.

Therefore m is a marker in $B_{\mathcal{G}}(u, d_{\mathcal{G}}(M, u) + i_1 + 2)$ and furthermore m is the smallest marker in $B_{\mathcal{G}}(u, d_{\mathcal{G}}(M, u) + i_2 - 2) \supseteq B_{\mathcal{G}}(u, d_{\mathcal{G}}(M, u) + i_1 + 2)$. We obtain $m(u, i) = m = m(v, i)$ for $i_1 + 2 \leq i \leq i_2 - 2$.

As this holds for any neighbour u, the node v cannot be a boundary node in the configurations $i_1 + 2 \leq i \leq i_2 - 2$. There are at most 4 configurations in the ends of the interval such that v may be a boundary node.   □

The algorithm FIND-CELLS is local. In the following sections, we use the cells and Lemma 2 to obtain local algorithms for the scheduling problems.

4.2 Sleep Scheduling

Let $\bar{C}(m, i) = \{v \in V_{\mathcal{G}} : d_{\mathcal{G}}(C(m, i), v) \leq 1\}$. For each marker m and configuration i, solve the LP

$$
\begin{aligned}
\text{maximise} \quad & \sum_K x_{m,i}(K) \\
\text{subject to} \quad & \sum_K K(v) x_{m,i}(K) \leq b(v) \quad \text{for all } v, \\
& x_{m,i}(K) \geq 0 \qquad \text{for all } K,
\end{aligned}
\tag{1}
$$

where v ranges over all nodes in $\bar{C}(m, i) \cap V_{\mathcal{R}}$, and K ranges over all subsets $K \subseteq \bar{C}(m, i) \cap V_{\mathcal{R}}$ such that K dominates $C(m, i) \cap V_{\mathcal{R}}$ in $\mathcal{R}$. Note that the boundary nodes may participate in domination, but they need not be dominated. The LP has constant size and depends on the local information only. Let $q_{m,i} = \sum_K x_{m,i}(K)$ be the total length of the solution.

Based on the computed solutions, each node controls its sleeping as follows. We use the synchronised clocks to proceed in cycles of length δ time units for some δ. Each cycle is further divided into $k\mu$ steps of length $\delta/(k\mu)$. We label the steps within each cycle by $0, 1, \ldots, k\mu - 1$. The behaviour of each node at step i is controlled as follows by the local solutions $x_{m,i}$ associated with the configuration i.

First, consider a non-boundary node $v \in V_{\mathcal{R}}$. The node constructs a schedule based on the solution $x_{m,i}$ where $m = m(v, i)$. All nodes in $\bar{C}(m, i) \cap V_{\mathcal{R}}$ consider the sets K with a nonzero $x_{m,i}(K)$ in the same order $K_1, K_2, \ldots$ (say, the lexicographic order). Let $t_j = \delta x_{m,i}(K_j)/(k\mu q_{m,i})$. First, if $v \in K_1$, the node is awake for t_1 time units; otherwise it is asleep for t_1 time units. Then, if $v \in K_2$, the node is awake for t_2 time units, and so on. This way we have scaled down the entire schedule $x_{m,i}$ into one time step of length $\delta/(k\mu)$.

Second, consider a boundary node $v \in V_{\mathcal{R}}$. As above, we construct a schedule based on $x_{m(v,i),i}$. Additionally we construct a schedule based on $x_{m(u,i),i}$ for every u such that $m(u, i) \neq m(v, i)$ and $\{u, v\} \in E_{\mathcal{G}}$. We take the union of these schedules: at any point in time, the node v is awake if it is awake according to at least one of the schedules.

In each configuration i, each node is a member of $C(m, i)$ for some m, and the local solution $x_{m,i}$ guarantees that $C(m, i) \cap V_{\mathcal{R}}$ is dominated at every point in time. Thus, this procedure is correct in the sense that $V_{\mathcal{R}}$ is dominated at every point in time, as long as no node runs out of battery.

Proof of Theorem 1. Let x be an optimal sleep schedule in the centralised setting; let $q = \sum_D x(D)$. The solution x can be used to construct a feasible solution to each local LP (1): for each dominating set D, add $x(D)$ units to $x_{m,i}(K)$ where $K = D \cap \bar{C}(m, i)$; as D dominates $V_{\mathcal{R}}$, it follows that K dominates $C(m, i) \cap V_{\mathcal{R}}$. Thus, $q_{m,i} \geq q$, as $x_{m,i}$ is an optimal solution to (1).

Consider an arbitrary node $v \in V_{\mathcal{R}}$. During each step i when v is not a boundary node, it is awake for at most $\delta b(v)/(k\mu q_{m,i}) \leq \delta b(v)/(k\mu q)$ time units. When v is a boundary node, this is increased by at most a factor $\Delta + 1$ because there are at most Δ neighbours and thus at most Δ different neighbouring cells. By Lemma 2, the node v is a boundary node in at most 4μ configurations out of $k\mu$. Thus, v is awake for at most $(k\mu + 4\Delta\mu)\delta b(v)/(k\mu q) = (1 + 4\Delta/k)\delta b(v)/q$ time units during an entire cycle of length δ. During $\lfloor q/(\delta(1 + 4\Delta/k)) \rfloor$ cycles, v is awake for at most $b(v)$ time units. Thus, the battery of v lasts at least $\lfloor q/(\delta(1 + 4\Delta/k)) \rfloor \delta \geq q/(1 + 4\Delta/k) - \delta$ time units. By choosing a small δ, we can obtain an approximation ratio $1 + \epsilon$ for any $\epsilon > 4\Delta/k$.

To choose a small enough δ, we need some information on q. If $q > 0$ then each node has to have at least one neighbour u with $b(u) > 0$; by letting all nodes be awake as long as their batteries last, we obtain a trivial constant lower bound for q from the smallest nonzero element of the finite set of possible $b(v)$; we use this bound to choose δ. The obtained δ (as well as any other value) is trivially valid also in the case $q = 0$. □

4.3 Activity Scheduling

Let $C^\circ(m, i)$ be the set of nodes $v \in C(m, i)$ such that v is not a boundary node in configuration i. For each marker m and configuration i, solve the LP

$$
\begin{aligned}
\text{minimise} \quad & \textstyle\sum_K x_{m,i}(K) \\
\text{subject to} \quad & \textstyle\sum_K K(v)x_{m,i}(K) \geq a(v) \quad \text{for all } v, \\
& x_{m,i}(K) \geq 0 \qquad\quad \text{for all } K,
\end{aligned}
\tag{2}
$$

where v ranges over all nodes in $C^\circ(m, i) \cap V_C$ and K ranges over all subsets $K \subseteq C^\circ(m, i) \cap V_C$ such that K is an independent set in V_C. Note that the boundary nodes are not considered. The LP has constant size and depends on the local information only. Let $q_{m,i} = \sum_K x_{m,i}(K)$ be the total length of the solution.

As in Sect. 4.2, we proceed in cycles of length δ and steps of length $\delta/(k\mu)$. Also the translation of local solutions into schedules is the same for nonboundary nodes. However, the boundary nodes are inactive.

Proof of Theorem 2. Let x be an optimal activity schedule in the centralised setting; let $q = \sum_D x(D)$. The solution x can be used to construct a feasible solution to each local LP (2): for each independent set I, add $x(I)$ units to $x_{m,i}(K)$ where $K = I \cap C^\circ(m, i)$; as I is an independent set in C, so is K. Thus, $q_{m,i} \leq q$, as $x_{m,i}$ is an optimal solution to (2).

Consider an arbitrary node $v \in V_C$. During each step i when v is not a boundary node, it is active for at least $\delta a(v)/(k\mu q_{m,i}) \geq \delta a(v)/(k\mu q)$ time units. By Lemma 2, the node v is a boundary node in at most 4μ configurations out of $k\mu$. Thus, v is active for at least $(1-4/k)\delta a(v)/q$ time units during an entire cycle of length δ. During $\lceil q/(\delta(1-4/k)) \rceil$ cycles, v is active for at least $a(v)$ time units. Thus, the node can complete its activities in $\lceil q/(\delta(1-4/k)) \rceil \delta \leq q/(1-4/k)+\delta$ time units. By choosing a small δ, we can obtain an approximation ratio $1/(1-\epsilon)$ for any $\epsilon > 4/k$. Again an appropriate δ can be chosen by bounding q using the information on the possible values of $a(v)$. □

4.4 A Lower Bound for Local Approximability

We proceed to show that the value ϵ in the approximation guarantees of Theorems 1 and 2 cannot be improved by a constant factor larger than 9.

Select integers $N \geq 100$ and $\mu \geq 10$. Consider an arbitrary local algorithm with local horizon $L \in \mathbb{N}$. Construct the communication graph $\mathcal{G}$ by forming a ring of $n = (6N + 1)6L$ nodes, that is, $V_\mathcal{G} = \{0, 1, \ldots, n - 1\}$, $E_\mathcal{G} = \{\{0, 1\}, \{1, 2\}, \ldots, \{n - 2, n - 1\}, \{n - 1, 0\}\}$. The identifiers are ordered by $0 < 1 < \ldots < n - 1$. Place the markers at the nodes v where $v \equiv 0 \pmod{6N + 1}$. The construction is a $(2, 3N, \lceil \mu(3N + 1/2) \rceil - 1, \mu)$-marked graph.

For sleep scheduling, let $\mathcal{R} = \mathcal{G}$ and $b(v) = 1$ for each $v \in V_\mathcal{R}$; for activity scheduling, let $\mathcal{C} = \mathcal{G}$ and $a(v) = 1$ for each $v \in V_C$. Consider the nodes $U = \{L, L+1, \ldots, n - L - 1\}$. For each $j \in \{0, 1, \ldots, 6N\}$, the local neighbourhoods of the nodes $v \in U$ with $v \equiv j \pmod{6N + 1}$ are identical; thus, each of these nodes must make the same decision at any point in time.

In the case of sleep scheduling, there exists a schedule of length 3. However, the local algorithm cannot achieve an optimal solution. Consider a chain of $6N + 1$ nodes in U. If only $2N$ nodes are awake at a given point in time, then only $6N$ nodes are awake in a chain of $18N + 3$ nodes, as each subchain of length $6N + 1$ behaves identically. However, $6N$ nodes cannot dominate a chain of $18N + 1$ nodes; thus, there is at least one node in the chain which cannot be dominated. Therefore at least $2N + 1$ nodes have to be awake, and the total lifetime of the nodes in a subchain of $6N+1$ nodes is thus at most $(6N+1)/(2N+1) = 3/(1+\epsilon)$ for $\epsilon = 2/(6N + 1) > 0.33/N$. Our local algorithm achieves the approximation guarantee $1 + \epsilon$ for any $\epsilon > 8/k$ where $k \geq 3N - 1/\mu - 3N/\mu - 1/2 \geq 2.694N$. That is, we can achieve $\epsilon = 9 \times 0.33/N$.

In the case of activity scheduling, there exists a schedule of length 2. In the local algorithm, for each chain of $6N + 1$ nodes there can be at most $3N$ nodes active simultaneously, implying that the length of the schedule obtained by the arbitrary local algorithm is at least $(6N+1)/(3N) = 2/(1-\epsilon)$ for $\epsilon = 1/(6N+1) > 0.165/N$. Our local algorithm achieves $\epsilon = 9 \times 0.165/N$.

In conclusion, we have presented an infinite family of parameters $(\Delta, \ell_1, \ell_\mu, \mu)$ such that the ϵ in our approximation guarantees for both sleep scheduling and activity scheduling is within factor 9 of the best possible that any deterministic local approximation algorithm can achieve.

In this lower bound, we focused on the case $\ell_1 \approx \ell_\mu/\mu \to \infty$. The following lemma shows that the case $\ell_1 \ll \ell_\mu/\mu$ is trivial to local algorithms.

Lemma 5. *If $\ell_\mu \geq (\mu+1)(\ell_1+1/2)$, then the size of each connected component of a $(\Delta, \ell_1, \ell_\mu, \mu)$-marked graph is bounded by a constant.*

Proof. Let $(\mathcal{G}, M)$ be a $(\Delta, \ell_1, \ell_\mu, \mu)$-marked graph with $\ell_\mu \geq (\mu + 1)(\ell_1 + 1/2)$. To reach a contradiction, assume that there exist nodes $v_0, v_\mu \in V_\mathcal{G}$ such that $d_\mathcal{G}(v_0, v_\mu) = \mu(2\ell_1 + 1)$. Then there exist nodes $v_1, v_2, \ldots, v_{\mu-1} \in V_\mathcal{G}$ such that $d_\mathcal{G}(v_i, v_{i+1}) = 2\ell_1 + 1$ and a node $u \in V_\mathcal{G}$ such that $d_\mathcal{G}(v_i, u) \leq \lceil \mu(\ell_1 + 1/2) \rceil$. For $i = 0, 1, \ldots, \mu$, let $m_i \in M$ be a marker having the minimum distance to v_i in $\mathcal{G}$; the nodes $m_0, m_1, \ldots, m_\mu$ are distinct. Furthermore, it holds that $d_\mathcal{G}(m_i, u) \leq d_\mathcal{G}(m_i, v_i) + d_\mathcal{G}(v_i, u) \leq \ell_1 + \lceil \mu(\ell_1+1/2) \rceil \leq \ell_1 + \mu(\ell_1+1/2) + 1/2 = (\mu + 1)(\ell_1 + 1/2) \leq \ell_\mu$. This implies that we have $\mu + 1$ markers in $B_\mathcal{G}(u, \ell_\mu)$, which is a contradiction with the assumption that $(\mathcal{G}, M)$ be a $(\Delta, \ell_1, \ell_\mu, \mu)$-marked graph. Thus, the diameter of each connected component of $\mathcal{G}$ is less than $\mu(2\ell_1+1)$, and each connected component consists of at most $1 + \Delta^{\mu(2\ell_1+1)}$ nodes. $\square$

5   Deploying a Marked Network

Any local algorithm for scheduling requires some auxiliary information, marking or otherwise, to break symmetry (Lemma 1). Thus, to apply a local algorithm, one must incorporate this information into the network when the network is deployed. In particular, a *practically feasible* way to deploy the network is required.

We conclude this paper by developing a series of examples that illustrate how one might go about and deploy a marked network in a physical area so that Definition 1 is met.

An intuitive picture to keep in mind in what follows is a graduate student walking about an area where a network is to be deployed with two (heavy) bags of sensor devices. One bag contains devices with the marker bit set, and the other bag devices with the bit reset.

We start with a purely combinatorial setup and proceed in steps towards more realistic scenarios.

5.1 Grids

Consider an infinite 2-dimensional grid graph $\mathcal{G}$, where $V_{\mathcal{G}} = \mathbb{Z}^2$ and $E_{\mathcal{G}} = \{\{(x_1, x_2), (y_1, y_2)\} : |x_1 - y_1| + |x_2 - y_2| = 1\}$. Choose an integer $N > 1$. Deploy the markers at nodes $M_{\mathcal{G}} = \{(Ni, Nj) : i + j \text{ odd}\}$. The constructed $(\mathcal{G}, M)$ is a $(4, N, 2N - 1, 4)$-marked graph. Generalisation to higher dimensions is immediate.

For large N we obtain an approximation ratio $1 + \epsilon$ where $\epsilon \approx 64/N$ for sleep scheduling and $\epsilon \approx 16/N$ for activity scheduling. In this sense, our local approximation algorithm is a *local approximation scheme* for grid graphs: any approximation ratio above 1 can be achieved by deploying the markers in a sufficiently sparse manner. Furthermore, the rule for deploying the markers is arguably practically feasible from the perspective of a combinatorial entity traversing the grid.

5.2 Globally Grid-Like Graphs

The communication topology in a real physical environment does not have a perfect grid structure. To arrive at a more versatile model, consider an infinite connected graph $\mathcal{H}$ where every node has degree at most $\Delta_{\mathcal{H}}$. We assume that the large-scale structure of $\mathcal{H}$ is similar to a 2-dimensional grid graph $\mathcal{G}$, but the small-scale structure of $\mathcal{H}$ can be arbitrary. In precise terms, we assume that the metric spaces $(V_{\mathcal{G}}, d_{\mathcal{G}})$ and $(V_{\mathcal{H}}, d_{\mathcal{H}})$ are *quasi-isometric*[1]; that is, we assume that there exist mappings $h: V_{\mathcal{G}} \to V_{\mathcal{H}}$, $g: V_{\mathcal{H}} \to V_{\mathcal{G}}$ and constants $C \geq 0$, $\lambda \geq 1$ such that $d_{\mathcal{H}}(h(x), h(y)) \leq \lambda d_{\mathcal{G}}(x, y) + C$, $d_{\mathcal{G}}(g(u), g(v)) \leq \lambda d_{\mathcal{H}}(u, v) + C$, $d_{\mathcal{G}}(g(h(x)), x) \leq C$, and $d_{\mathcal{H}}(h(g(v)), v) \leq C$ for all $x, y \in V_{\mathcal{G}}$ and $u, v \in V_{\mathcal{H}}$. Define the marking of $\mathcal{H}$ from a marking of $\mathcal{G}$ by $M_{\mathcal{H}} = h(M_{\mathcal{G}})$.

Lemma 6. *Any marked graph $(\mathcal{H}, M_{\mathcal{H}})$ that satisfies the above conditions is a $(\Delta_{\mathcal{H}}, \lfloor \lambda N + 2C \rfloor, \lfloor 2\lambda N - (2C + 1)/\lambda \rfloor, \lceil 2\lambda^2 \rceil^2)$-marked graph where N is the constant used to mark $\mathcal{G}$.*

Proof. Let $v \in V_{\mathcal{H}}$. Let m be the marker closest to $g(v)$ in $\mathcal{G}$; $d_{\mathcal{G}}(m, g(v)) \leq N$. We have $h(m) \in M_{\mathcal{H}}$ and $d_{\mathcal{H}}(h(m), v) \leq d_{\mathcal{H}}(h(m), h(g(v))) + d_{\mathcal{H}}(h(g(v)), v) \leq \lfloor \lambda d_{\mathcal{G}}(m, g(v)) + C \rfloor + \lfloor C \rfloor \leq \lfloor \lambda N + C \rfloor + \lfloor C \rfloor \leq \lfloor \lambda N + 2C \rfloor$. We can choose $\ell_1 = \lfloor \lambda N + 2C \rfloor$.

---

[1] Ghys [19] attributes this definition of quasi-isometry to G. A. Margulis.

Let $\ell_\mu = \lfloor 2\lambda N - (2C+1)/\lambda \rfloor$. Let $v \in V_{\mathcal{H}}$ and $m \in M_{\mathcal{G}}$ be such that $d_{\mathcal{H}}(v, h(m)) \leq \ell_\mu$. Then $d_{\mathcal{G}}(g(v), m) \leq d_{\mathcal{G}}(g(v), g(h(m))) + d_{\mathcal{G}}(g(h(m)), m) \leq \lambda \ell_\mu + 2C \leq 2\lambda^2 N - 1 \leq \lceil 2\lambda^2 \rceil N - 1$. For any $x \in V_{\mathcal{G}}$ and positive integer κ there are at most κ^2 markers in $B_{\mathcal{G}}(x, \kappa N - 1)$; thus there are at most $\lceil 2\lambda^2 \rceil^2$ markers in $B_{\mathcal{H}}(v, \ell_\mu)$. We can choose $\mu = \lceil 2\lambda^2 \rceil^2$. $\qquad\square$

Again we obtain an approximation scheme; any approximation ratio above 1 can be achieved by placing the markers in a sufficiently sparse manner in $\mathcal{G}$.

Intuitively, each element of $V_{\mathcal{G}}$ corresponds to a geometric area and $g(v) \in V_{\mathcal{G}}$ is the area where the device $v \in V_{\mathcal{H}}$ is located. The choice of $M_{\mathcal{H}}$ reflects the following rule for deploying the markers. First, some geometric areas $M_{\mathcal{G}}$ are selected based on the grid structure. Second, one marker is deployed in each of these geometric areas.

In the small scale, quasi-isometry allows arbitrary structure to occur; the small-scale structure of realistic communication graphs is irregular and unpredictable due to the complex nature of the physics of radio propagation. In the large scale, quasi-isometry requires that shortest-path distances in the communication graph reflect the distances in the ambient space; this is a reasonable assumption from a dense deployment of sensors in an area devoid of large-scale obstructions.

5.3   Serendipity of Locality

The defining property of a local algorithm is that the behaviour of each network node is uniquely determined by the radius-L neighbourhood of the node. In other words, all things being equal in the neighbourhood, the large-scale topology of the network has no effect in the operation of a network node. This is particularly useful from the perspective of network deployment—to fulfil the intended sensing objective, it suffices to deploy the sensor nodes in a manner that, from the perspective of mission-critical sensor nodes, *looks like* a benign topology, even if the actual topology is not.

In more concrete terms, let us assume that we have some two-dimensional area $\mathcal{A}$ of arbitrary shape that we want to monitor by a sensor network. Let us also assume that we have a method of network deployment that would produce a $(\Delta, \ell_1, \ell_\mu, \mu)$-marked graph if applied to the infinite two-dimensional plane; say, the method produces a globally grid-like marked graph $(\mathcal{H}, M_{\mathcal{H}})$.

Now, to deploy a network to monitor $\mathcal{A}$, all one has to do is to apply the deployment method to $\mathcal{A}$ plus its constant-width surroundings. More precisely, we deploy so that for any node v that is placed within $\mathcal{A}$, its $(L+1)$-hop neighbourhood is indistinguishable from its neighbourhood in $(\mathcal{H}, M_{\mathcal{H}})$. By locality it follows immediately that any node within $\mathcal{A}$ (or with a neighbour in $\mathcal{A}$) operates exactly as it would operate in the case of $(\mathcal{H}, M_{\mathcal{H}})$. For example, if the nodes execute the sleep scheduling algorithm, then full coverage for every node within $\mathcal{A}$ is guaranteed, with a lifetime at least as good as in the case of an infinite graph. Other nodes may fail in an arbitrary manner, but this does not affect the operation within $\mathcal{A}$; for example, in the case of activity scheduling, these nodes cannot conflict with the nodes within $\mathcal{A}$.

Again it can be argued that this deployment scheme is straightforward to implement in practice. The overhead (in the number of extra nodes that need to be deployed outside $\mathcal{A}$) depends on the shape and size of $\mathcal{A}$, but if the shape of $\mathcal{A}$ is not too irregular, the relative overhead approaches zero as the surface area of $\mathcal{A}$ increases.

Both the deployment of the markers and the deployment of the extra nodes in the surroundings of $\mathcal{A}$ can be seen as examples of a basic tradeoff in computational effort: minor (and, from the perspective of the deployer, computationally straightforward) extra effort invested in deployment pays off by enabling local approximation of fundamental scheduling problems.

Acknowledgements. This research was supported in part by the Academy of Finland, Grants 116547 and 117499, and by Helsinki Graduate School in Computer Science and Engineering (Hecse).

References

1. Naor, M., Stockmeyer, L.: What can be computed locally? SIAM Journal on Computing 24(6), 1259–1277 (1995)
2. Cardei, M., MacCallum, D., Cheng, M.X., Min, M., Jia, X., Li, D., Du, D.Z.: Wireless sensor networks with energy efficient organization. Journal of Interconnection Networks 3(3–4), 213–229 (2002)
3. Floréen, P., Kaski, P., Suomela, J.: A distributed approximation scheme for sleep scheduling in sensor networks. In: SECON 2007. Proc. 4th Annual IEEE Communications Society Conference on Sensor, Mesh and Ad Hoc Communications and Networks, San Diego, CA, June 2007, pp. 152–161. IEEE, Piscataway (2007)
4. Koushanfar, F., Taft, N., Potkonjak, M.: Sleeping coordination for comprehensive sensing using isotonic regression and domatic partitions. In: Proc. 25th Conference on Computer Communications (INFOCOM, Barcelona, April 2006), IEEE, Piscataway, NJ (2006)
5. Moscibroda, T., Wattenhofer, R.: Maximizing the lifetime of dominating sets. In: IPDPS 2005. Proc. 19th IEEE International Parallel and Distributed Processing Symposium, Denver, CO, IEEE Computer Society Press, Los Alamitos (2005)
6. Pemmaraju, S.V., Pirwani, I.A.: Energy conservation via domatic partitions. In: MobiHoc 2006. Proc. 7th ACM International Symposium on Mobile Ad Hoc Networking and Computing, Florence, pp. 143–154. ACM Press, New York (2006)
7. Jain, K., Padhye, J., Padmanabhan, V.N., Qiu, L.: Impact of interference on multihop wireless network performance. Wireless Networks 11(4), 471–487 (2005)
8. Feige, U., Halldórsson, M.M., Kortsarz, G., Srinivasan, A.: Approximating the domatic number. SIAM Journal on Computing 32(1), 172–195 (2002)
9. Lund, C., Yannakakis, M.: On the hardness of approximating minimization problems. Journal of the ACM 41(5), 960–981 (1994)
10. Suomela, J.: Locality helps sleep scheduling. In: WSW 2006. Working Notes of the Workshop on World-Sensor-Web: Mobile Device-Centric Sensory Networks and Applications, Boulder, CO, October 2006, pp. 41–44 (2006), http://www.sensorplanet.org/wsw2006/
11. Linial, N.: Locality in distributed graph algorithms. SIAM Journal on Computing 21(1), 193–201 (1992)

12. Kuhn, F., Moscibroda, T., Wattenhofer, R.: The price of being near-sighted. In: SODA 2006. Proc. 17th Annual ACM-SIAM Symposium on Discrete Algorithm, Miami, FL, January 2006, pp. 980–989. ACM Press, New York (2006)
13. Hochbaum, D.S., Maass, W.: Approximation schemes for covering and packing problems in image processing and VLSI. Journal of the ACM 32(1), 130–136 (1985)
14. Erlebach, T., Jansen, K., Seidel, E.: Polynomial-time approximation schemes for geometric intersection graphs. SIAM Journal on Computing 34(6), 1302–1323 (2005)
15. Hunt III, H.B., Marathe, M.V., Radhakrishnan, V., Ravi, S.S., Rosenkrantz, D.J., Stearns, R.E.: NC-approximation schemes for NP- and PSPACE-hard problems for geometric graphs. Journal of Algorithms 26(2), 238–274 (1998)
16. Jiang, T., Wang, L.: An approximation scheme for some Steiner tree problems in the plane. In: Du, D.-Z., Zhang, X.-S. (eds.) ISAAC 1994. LNCS, vol. 834, pp. 414–422. Springer, Heidelberg (1994)
17. Suomela, J.: Approximability of identifying codes and locating-dominating codes. Information Processing Letters 103(1), 28–33 (2007)
18. Kuhn, F., Nieberg, T., Moscibroda, T., Wattenhofer, R.: Local approximation schemes for ad hoc and sensor networks. In: DIALM-POMC 2005. Proc. Joint Workshop on Foundations of Mobile Computing, Cologne, September 2005, pp. 97–103. ACM Press, New York (2005)
19. Ghys, É.: Les groupes hyperboliques. Astérisque 189–190, 203–238 (1990), [Séminaire Bourbaki, vol. 1989/90, Exp. No. 722]

Assigning Sensors to Missions with Demands[*]

Amotz Bar-Noy[1,2], Theodore Brown[1,3], Matthew P. Johnson[1],
Thomas La Porta[4], Ou Liu[1], and Hosam Rowaihy[4]

[1] The Graduate Center of the City University of New York
[2] Brooklyn College, City University of New York
[3] Queens College, City University of New York
[4] Pennsylvania State University

Abstract. We introduce SEMI-MATCHING WITH DEMANDS (SMD),
which models a certain problem in sensor networks of assigning individual sensors to sensing tasks. If there are multiple sensing tasks or
missions to be accomplished simultaneously, and if sensor assignment
must be exclusive, then this is a bipartite semi-matching problem. Each
mission is associated with a demand value and a profit value; each sensor-
mission pair is associated with a utility offer (possibly 0). The goal is a
sensor assignment that maximizes the profits of the satisfied missions
(with no credit for partially satisfied missions). SMD is **NP**-hard and
as hard to approximate as MAXIMUM INDEPENDENT SET. Therefore we
investigate less difficult constrained versions of the problem. We give a
simple greedy Δ-approximation algorithm for a degree-constrained version (Δ-SMD), in which each mission receives positive utility offers from
at most Δ sensors. For small Δ, we show that Δ-SMD is equivalent to
k-SET PACKING (with $k = \Delta$), which yields a polynomial-time $(\Delta+1)/2$-
approximation. For $\Delta = 2$, we solve the problem optimally by reduction
to maximum matching. Finally, we introduce a geometric version which
remains strongly **NP**-hard but has a PTAS.

1 Introduction

A sensor network consists of a large number of small sensing devices that are
able to collect information about their surroundings. When a sensor network is
deployed in the field it may be tasked with achieving multiple, possibly conflicting, missions. Hence, schemes that match sensor resources to mission demands
become necessary. If there are multiple sensors and multiple missions, we must

[*] Research was sponsored by the U.S. Army Research Laboratory and the U.K.
Ministry of Defence and was accomplished under Agreement Number W911NF-06-
3-0001. The views and conclusions contained in this document are those of the
author(s) and should not be interpreted as representing the official policies, either
expressed or implied, of the U.S. Army Research Laboratory, the U.S. Government,
the U.K. Ministry of Defence or the U.K. Government. The U.S. and U.K. Govern-
ments are authorized to reproduce and distribute reprints for Government purposes
notwithstanding any copyright notation hereon. We thank Panos Hilaris for deliver-
ing the workshop presentation.

M. Kutyłowski et al. (Eds.): ALGOSENSORS 2007, LNCS 4837, pp. 114–125, 2008.

choose the best matching of sensors to missions. A given sensor may offer different missions varying amounts of information (because of geometry, obstructions, or remaining battery level, for example), or none at all.

Missions may vary in both importance (*profit*) and difficulty (*demand*), and these properties need not be correlated. An ongoing surveillance mission may be expensive but of minor importance, whereas an urgent mission for information about one particular spot may be low-demand but very important. In many applications, partial satisfaction will be no better than zero satisfaction. If the goal of a given mission is to reconstruct the 3D shape of an object, for example, then this may be accomplished with images from two cameras, but an image from just one camera will be useless. Indeed, accepting the single image could actually be harmful since the drain on the sensor's battery could preclude a future mission that might otherwise have been satisfiable. In our model we only receive profit, therefore, from missions whose demands are fully met. Hence the problem is to choose the "best" assignment of sensors to missions, in the sense that profits from satisfied missions are maximized.

Since this problem is **NP**-hard even to approximate, we investigate constrained versions for which approximation algorithms exist. First, we bound the number of sensors that may offer contributions to any single mission. This is a reasonable assumption in realistic settings in which sensors have a limited sensing range and the sensors are distributed in such a way as to limit sensing redundancy. Indeed, covering an entire field using as few sensors as possible is an important problem in sensor networks (see Section 2). Second, we assume that sensors and missions are located at points in the plane or higher-dimensional metric space; we also assume a bounded sensing range and a bounded density of sensors and missions. These assumptions may apply to settings in which sensors are physical objects that exist in the world and that take up space and missions can be similarly localized.

The rest of this paper is organized as follows. Section 2 discusses some related work in sensor networks and in assignment problems. In Section 3 we formally define the sensor assignment problem and study its computational complexity. In Section 4, we introduce the degree-bounded version. We give a Δ-approximation greedy algorithm for degree Δ and an efficient optimal algorithm for degree 2. We also show that the problem is equivalent to Δ-SET PACKING, which yields a $(\Delta + 1)/2$-approximation. In Section 5 we introduce a geometric version and give a shifting-based PTAS. Finally, Section 6 concludes the paper.

2   Related Work

Sensor networks. The general problem of choosing sensors to achieve an objective has received sizable attention lately. Several selection objectives have been considered. In [20,22], for example, the goal is to cover the region using few sensors in order to conserve energy. Sensor selection schemes have also been proposed to efficiently track and locate targets. In [24], for example, the most informative sensor for tracking a target is chosen based on the concept of information gain. This information is then passed on to the next active node, which is chosen by considering the target's expected path. Target localization using acoustic sensors is considered by [14]. The goal there is to minimize the mean squared error of the target location as perceived by the active nodes. Our work, however, is motivated by contention between multiple missions with varying profit values, and therefore focuses on mission selection rather than sensor selection.

There has also been some work on frameworks for single and multiple mission assignment problems. [3] e.g. defines a framework modeling the assignment problem with notions of utility and cost. The goal is to find a solution that maximizes the utility while staying under a predefined budget. In [18], a market-based method is used in which sensors provide information or "goods" which can be purchased while observing certain budgets. Our problem differs in two important respects. First, we maximize the profits uncategorically; the only "budget" is the set of available sensors. Second, we do not simply maximize total received utility. Solution quality for our problem is more strict in that partial satisfaction receives no partial credit.

Algorithms. Although we use the terminology of sensors and missions for concreteness, SMD can be viewed as a more general problem of resource allocation. An alternative interpretation regards scheduling jobs on unrelated parallel machines. As in other (maximization) scheduling problems [23], the goal is a schedule that maximizes profit earned from jobs completed, subject to certain constraints. The twist is that each job specifies not the set of *machines* that can perform it, but the set of *families of machines* that can perform it. (A job may be too difficult to be performed by any single machine.) The feasibility constraint is that no machine can be assigned more than one "sub-job".

SMD also relates to other optimization problems, such as the BIN *Covering* problem, in which the goal is to use a set of items to fill completely as many bins as possible. SMD is a generalization of (weighted) Bin Covering in that an "item" may take up different amounts of space in different "bins". In this way, an analogy can be seen between BIN COVERING and SMD and the MULTIPLE KNAPSACK and GENERALIZED ASSIGNMENT [7] problems.

Combinatorial Auctions. In contrast to conventional auctions, in combinatorial auctions players can bid of sets or *combinations* of items (which may entail exponentially many bids). Given a fixed supply of goods, the goal of the winner determination problem is to maximize revenue earned from the sale of disjoint item combinations. Since this is a difficult problem, much of the research focus has been on AI or algorithm-engineering approaches. (See [5] for a survey.) Another way of understanding our problem is as a combinatorial auction in which bidders are the missions, the items are the sensors *and* the missions, and each mission places a bid (equal to its own profit) for any set that can be constructed as follows: the set contains the mission itself and some subset of sensors that together satisfy the mission's demand. The language of profits, demands, and edge values, however, allows for a succinct representation of bids.

Many weaker and often fractional models of sensor-mission assignment can be reduced to maximal matching or network flow problems, and thus can be solved optimally in polynomial time [1]. A survey of sensor selection and assignment problems, including simple theoretical models of the problem, can be found in [21].

3   Problem Definition

Given is a complete weighted bipartite graph, whose vertex sets consist of sensors $S = \{S_1, ..., S_n\}$ and missions $\{M_1, ..., M_m\}$. A positively weighted edge (S_i, M_j) means that S_i is applicable to M_j. The weight of the edge (e_{ij}) indicates the utility (or quality of information) that S_i could contribute to M_j if this pairing were chosen. Also given is a positive-valued demand d_j associated with each mission M_j, indicating the total utility the mission requires. What we seek is a *semi-matching* of sensors to missions, so that (ideally) each mission demand is satisfied. That is, a sensor may be assigned to at most one of the missions to which it is applicable, but it is legal for a mission to accept utility from multiple sensors. Of course, satisfying all missions may not be feasible; in general, the goal is to maximize a weighted sum of the *satisfied missions*. We assume there is a profit p_j associated with achieving mission M_j. We then seek to maximize the total satisfied profit. Note that there is no profit awarded for a partially satisfied mission in this model.

Unless there is structure to the weights of the sensor-mission edges, for example if they relate to the geometry of node positions, we can assume without loss of generality that each demand is 1. For each mission M_j with demand d_j, simply divide edge value e_{ij} by d_j to obtain an instance with unit demands. Unless otherwise stated, we will assume this normalization henceforth, though it is sometimes convenient to allow for non-unit demands. With this in mind, we define the problem formally.

Instance: A weighted bipartite graph $G = (S, M, P, E)$, where $S = \{S_1, ..., S_n\}$ is a collection of sensors, $M = \{M_1, ..., M_m\}$ is a collection of missions, $P = \{p_1, ..., p_m\}$ is a collection of positive mission profits, and E is a collection of non-negative weights for the edges $S \times M$.

Goal: Find a semi-matching $F \subseteq E$ (no two chosen edges share the same *sensor*), in which $\sum_{M_j \in A} p_j$ is maximized, where $A \subseteq M$ is the set of missions satisfied by F (i.e., $\sum_{(i,j) \in F} e_{ij} \geq 1$ for each $M_j \in A$).

The problem can easily be expressed by an Integer Programming formulation. The formulation below employs two sets of decision variables: y_j indicating whether mission M_j is satisfied, and x_{ij} indicating whether sensor S_i is assigned to mission M_j. Finding a solution can be seen as a two-step process: decide which missions to satisfy, and then decide how to satisfy them. Each mission M_j has a constraint requiring that the sum of utility received by M_j be at least the value y_j, which is 0 or 1. When $y_j = 0$, this constraint is automatically satisfied.

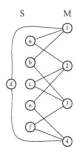

Fig. 1. Integrality gap instance

Maximize: $\sum_j p_j y_j$
Such that: $\sum_{i=1}^{n} x_{ij} e_{ij} \geq y_j$, for each mission M_j,
$\qquad \sum_{j=1}^{m} x_{ij} \leq 1$, for each sensor S_i, and
$\qquad x_{ij} \in \{0,1\}$, for each variable x_{ij} and $y_j \in \{0,1\}$, for each variable y_j

Note that if we had not normalized to unit demands, the first constraint would be: $\sum_{i=1}^{n} x_{ij} e_{ij} \geq y_j d_j$.

Remark 1. This IP has unbounded integrality gap, since instances can be constructed in which $OPT_{LP} = m/2$ and $OPT_{IP} = 1$, where m is the number of missions. To create such an instance (see Fig. 1), introduce and connect a separate sensor to each *pair of missions*, so that each mission has $m - 1$ neighbors, and set all demands to $m-1$ and all profits to 1. Then setting all edge weights to $1/2$ will clearly half-satisfy each mission, but only one can be satisfied integrally.

A relaxed version of this problem, in which profits are awarded fractionally for partial satisfaction *and* sensors can be assigned fractionally to multiple missions, can be solved with this formulation by Linear Programming, and certain versions can be solved by reduction to max-flow (see [21]), but this version cannot. A well known reduction from MIS [15] for the winner determination problem for Combinatorial Auctions [15] also applies to the (more specific) SMD problem. We briefly sketch it for completeness.

Proposition 1. *SMD is **NP**-hard and at least as hard to approximate as* MAX-IMUM INDEPENDENT SET *(MIS).*

Proof. Given an MIS graph $G = (V, E)$, an SMD instance is created with a mission M_v for each each $v \in V$, with $d_v = deg(v)$ and $p_v = 1$, and a sensor $S_{u,v}$ for each edge $(u, v) \in E$, which offers utility 1 to missions M_u and M_v. Then the optimal SMD solution yields the the optimal maximum independent set (and both share the same solution quality). QED

Because of this reduction, known hardness results for MIS also apply to SMD. MIS is the same as MAXIMUM CLIQUE on the complement graph, and MAXIMUM CLIQUE is known to be hard to approximate within $|V|^{1-\epsilon}$ for any $\epsilon > 0$, unless

---

Algorithm 1. Δ-Approximation Greedy Algorithm

---

for each mission M_j in order of decreasing P_j
    for each *still-available* sensor S_i in order of decreasing e_{ij}
        assign S_i to M_j
        if M_j is satisfied **then break**
    if M_j is not satisfied **then**
        return any sensors assigned to it

NP=ZPP (and hard within $|V|^{1/2-\epsilon}$ even without this assumption) [9]. This means that the best achievable approximation ratio can be little better than m, i.e., satisfying *only about one mission out of m*, which can be done by simple greedy algorithms.

We briefly note two special cases of this problem. The hardness properties above apply to both special cases.

Profits = demands: Set p_j to the original (pre-normalization) demand d_j.
Cardinality: Set $p_j = 1$, in which case the objective function is simply the number of satisfied missions.

4 Degree-Bounded Approximation Problem

Because of the difficulty of the approximation problem as defined, we will constrain it in order to render it more tractable. Let $|OPT|$ indicate the optimal solution value for a given problem instance, and let $|ALG|$ indicate a corresponding approximate solution value. We will say that an algorithm is a *c-approximation* for $c \geq 1$ if $c \leq \frac{|OPT|}{|ALG|}$ for every problem instance.

We will assume that the problem instance has *bounded degree*, in the following sense. If a sensor S_i makes a non-zero offer to a mission M_j, then say that S_i is M_j's *neighbor*. Then the assumption is that no mission has more than Δ neighbors, for some small constant Δ. (If all zero-weight edges are removed, this is the same as saying Δ bounds the degrees of all mission nodes in the SMD graph.) We call this problem Δ-SMD.

A simple greedy algorithm considers missions in decreasing order of profit. For each mission, the algorithm assigns it available sensors in decreasing order of offer utility, until the mission is satisfied. If the mission does not succeed, then all sensors are returned. Assuming that $m = O(n \log n)$, the running time is $O(mn \log n)$.

Definition 1. *Let a* star *consist of a mission and a* minimally satisfying *set of sensors for it. The sensor set is minimal in the sense that no proper subset of it would completely satisfy the mission in question. (Notice that a given mission may in general be part of many stars.) Say that a mission is* tight *if it has degree Δ and requires all Δ sensors in order to be satisfied. Two stars* overlap *if they share one or more sensors, if they share a mission, or both, including the case that the stars are identical.*

Proposition 2. *Algorithm 1 produces a Δ-approximation.*

Proof. Let OPT be the set of missions satisfied in some optimal solution (with solution quality $|OPT|$), and let ALG be the missions satisfied by Algorithm 1 (with quality $|ALG|$). We want to show that $|OPT| \leq \Delta \cdot |ALG|$, i.e., that

$$\sum_{M_j \in OPT} p_j \leq \sum_{M_{j'} \in ALG} \Delta \cdot p_{j'} \tag{1}$$

To prove Ineq. 1, we account for each term p_j on the LHS with one of the terms $\Delta \cdot p_{j'}$ on the RHS. For each $M_j \in OPT$, say that M_j *charges to* the highest-profit mission $M_{j'} \in ALG$ whose star overlaps with M_j's star, and write $M_j \in ch(M_{j'})$. (There must be one such $M_{j'}$ with $p_j \leq p_{j'}$ since Algorithm 1 satisfies a maximal set of missions, selected in decreasing order of profit.) Then let $M_{j'}$ be an arbitrary mission in ALG. $M_{j'}$ is either tight or not. Suppose tight, in which case that mission has only one star. If $M_{j'} \in OPT$, then only $M_{j'}$ itself charges to $M_{j'} \in ALG$; if $M_{j'} \notin OPT$, then at most Δ stars in OPT can charge to $M_{j'}$ (those that share at least one of its sensors). Now suppose $M_{j'}$ is not tight, so that it contains $\leq \Delta - 1$ sensors. Then at most Δ stars in OPT can charge to $M_{j'}$ (those that share at least one of its sensors, and possibly one that shares its mission). Thus we have

$$\sum_{M_j \in ch(M_j')} p_j \leq \Delta \cdot p_{j'} \tag{2}$$

By summing Ineq. 2 over all missions in ALG, we obtain Ineq. 1. QED

It is easy to construct an example with $\Delta + 1$ missions to show that the Δ bound is tight: let one mission have profit $1 + \epsilon$ and require all Δ sensors; let the rest have profit 1 and require one sensor each.

Corollary 1. *By the* MIS *reduction, Δ-SMD (for $\Delta \geq 3$) is **APX**-hard [19]. Given the approximation of Algorithm 1, Δ-SMD (for $\Delta \geq 3$) is **APX**-complete.*

Proposition 3. *2-SMD is in* **P**.

Proof. We reduce to the (weighted) maximal matching problem (see Fig. 2). The node set for the resulting graph will consist of the 2-SMD instance's sensors and missions. Whenever a mission M_j will be satisfied only by both its neighbors S_{i1}, S_{i2}, draw an edge (S_{i1}, S_{i2}) with weight equal to the mission profit; whenever a mission M_j will be satisfied by a single sensor S_i, draw an edge (S_i, M_j) with weight equal to the mission profit. Now find a maximal weighted matching in this (non-bipartite) graph in polynomial time. Each selected edge corresponds to a satisfied mission. It is clear that no sensor or mission will be used more than once. The optimal solution values of the matching graph and the SMD are by construction the same. QED

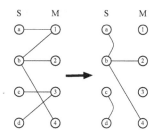

Fig. 2. Converting SMD to graph

Since the graph of a 2-SMD instance is sparse, the maximal weighted matching can be found in time $O(m^2 \log m)$ [8], where m is the number of missions. The time for finding the maximal matching is the dominant component of the running time.

We now relate Δ-SMD and Δ-SET PACKING. These problems turn out to be equivalent for Δ "small enough". In the SET PACKING problem, we are given a family of subsets of a universe of elements. Each subset has a positive weight. The goal is to choose a max-weight family of subsets without using any element more than once. Δ-SET PACKING is the variant of SET PACKING in which each set has at most Δ elements.[1]

Proposition 4. Δ-SMD *reduces to* Δ-SET PACKING, *when* $\Delta = O(\log nm)$.

Proof. The idea of the reduction is that each star in our SMD instance will become a set in the SET PACKING instance. (Since a given mission may have degree Δ, it can have $O(2^\Delta)$ many stars. Because of the bound on Δ, however, the resulting SET PACKING instance will be at most polynomially larger than the initial SMD instance size.) Specifically, a star with $s < \Delta$ sensors will become an $s+1$-element set containing the star's sensors and mission; a star with Δ sensors will become a Δ-element set containing only the star's sensors. Since a mission can have only one tight star of size Δ, the mission need not be included in the resulting star. Choosing a max-weight family of disjoint sets will now be the same as choosing a max-weight set of disjoint stars. It is interesting to note that the distinction between sensors and missions has now disappeared: no element can be used more than once. QED

Proposition 5. Δ-SET PACKING *reduces to* Δ-SMD.

Proof. Each element in the Δ-SET PACKING universe will correspond to a sensor in the resulting Δ-SMD instance. For each set in the Δ-SET PACKING instance, we create a separate mission that requires all the sensors in this set in order to be satisfied. Then each mission has degree at most Δ by construction. QED

For small enough Δ relative to problem instance size, Δ-SET PACKING and Δ-SMD are equivalent. Therefore for such Δ, Δ-SMD can be construed both a covering problem and a packing problem. An existing local search algorithm

---

[1] The parameter used is typically k, but we are interested in the case $k = \Delta$.

from Berman [2] gives a $(\frac{\Delta+1}{2}+\epsilon)$-approximation for $\Delta+1$-CLAW-FREE MIS, which Δ-SET PACKING reduces to. (ϵ is a running-time parameter, specifically, a $\frac{k}{(k-1)}\frac{\Delta+1}{2}$ approximation can be found in time polynomial in $O(kn)$.) Hence there is a $\frac{\Delta+1}{2}$ approximation for Δ-SMD (for small Δ). It was recently shown [10] that even for the cardinality version, approximating Δ-SET PACKING within a factor better than $\frac{\Delta}{\ln\Delta}$ is NP-hard. Following Jain et al.'s LP for FACILITY LOCATION [13], we can define a simpler IP formulation in terms of stars. When $\Delta = O(\log nm)$, there will only be polynomially many stars for a single problem instance. For a given element A (either sensor or mission), there will certainly be at most polynomially many stars containing A. In the following IP, decision variable y_t indicates that we choose star t; we have a constraint for each element. Intuitively, this IP has the advantage that it has just one set of decisions to make: which stars to pick? The profit for a star is simply the profit for the mission it includes; each sensor or mission can be used at most once.

Maximize: $\sum_t p_t y_t$
Such that: $\sum_{R_t:A_s \in R_t} y_t \leq 1$, for each sensor or mission A_s and
    $y_t \in \{0,1\}$, for each variable y_t

Remark 2. This IP has integrality gap at least $\frac{\Delta+1}{2}$, since instances can be constructed in which $OPT_{LP} = \frac{\Delta+1}{2}$ and $OPT_{IP} = 1$, where m is the number of missions. In fact, $\frac{\Delta+1}{2}$ is also a lower bound on the integrality gap of the first IP formulation, in the case of bounded degree.

5   Geometric Approximation Problem

We now introduce GEOMETRIC SMD (or GEOMSMD), in which all nodes (i.e., sensors and missions) lie in the plane (or higher-dimensional space) and geometrically inspired constraints are imposed. First, each sensor and mission now lie at a particular point in the plane. Second, we assume sensors have a bounded sensing range, i.e., e_{ij} can only be non-zero when the distance between S_i and M_j is less than this bound. (Without loss of generality, let the sensing range be 1. In this case, every star will lie in a unit disk.) We also assume a geometric analog to bounded degree, specifically an upper bound on the number of sensors or missions contained in any unit disk. This constraint will be satisfied automatically if the graph is *drawn in a civilized manner* [12], i.e, so that any two nodes are separated by some minimum distance $\lambda > 0$. Hence GEOMSMD is a special case Δ-SMD for some Δ.

The NP-hardness argument will involve the UNIT-DISK MIS (UD-MIS) problem [4], which is MIS variant in which the problem instance is the *intersection graph* for a set of unit disks lying in the plane. Equivalently, UD-MIS can be defined so that given a set of points in the plane, two points are connected by an edge iff their distance is strictly less than a global constant. We will argue that the NP-hardness proof for UD-MIS also applies to a density-bounded UD-MIS. The NP-hardness proof for UD-MIS from [4] is recounted in [16].

Algorithm 2. Shifting PTAS (error ϵ)

$k \leftarrow \lceil 2/\epsilon \rceil; \quad S = \emptyset$
for each $(i,j) \in [0,k)^2$
    lay the mesh with offset $(i,j); \quad S_{ij} \leftarrow \emptyset$
    for each cell C_t within the mesh
        $S_{ij} \leftarrow S_{ij} \cup opt(C_t)$
    if $val(S) < val(S_{ij})$ **then** $S \leftarrow S_{ij}$

Proposition 6. GEOMSMD *is strongly **NP**-hard.*

Proof. We reduce from 3SAT to UD-MIS [4,16] to GEOMSMD. Given the 3SAT instance, first apply the known UD-MIS reduction, which results in a UD-MIS graph and a number k (there is an independent set of size k iff the formula was satisfiable). It is clear that the resulting UD-MIS instance can be drawn with at most $O(1)$ disk per unit square (by inspection, at most 4 disks intersect at any point, and squeezing a chain of 3 disks so that all 3 centers lie in a unit disk will introduce a new edge). There are $O(\#vars \cdot \#clauses)$ disks, which is clearly polynomial in the 3SAT size.

Now we convert this UD-MIS decision-problem instance (G, k) into a GE-OMSMD decision-problem instance (G', k), by replacing each disk with a mission at the disk's center, and every maximal intersection of disks with a sensor needed by all of them. Since each mission needs all the sensors lying in its disk in order to be satisfied, k missions can be satisfied iff k independent disks can be chosen. Since in the UD-MIS construction each unit square contains at most $O(1)$ such intersections, in the resulting GEOMSMD instance, each unit square will contain at most $O(1)$ missions and $O(1)$ sensors, and the sensing distance is respected by construction. Thus 3SAT is reduced to GEOMSMD. QED

Since 3SAT is strongly **NP**-hard, it follows that an FPTAS is unlikely for GE-OMSMD. A PTAS, which we now give, is the best that can be hoped for. We employ the shifting technique originally introduced by Hochbaum & Maass [11]. Within a $c \times c$ cell, there will be at most $O(\Delta c^2)$ sensors and missions, for some constant Δ. As c increases, the fraction of the a cell's area near to the edge will decrease. For a $c \times c$ cell, the internal portion can be solved brute-force in time exponential in c and Δ, but polynomial in the problem instance size nm.

We now give the PTAS,[2] which is similar to the UNIT-DISK MIS [17] PTAS (following the presentation in [6]). For now, assume for simplicity that all points are bounded by a square region I whose size is polynomial in the input size. Then for a desired error bound ϵ, we can choose $c = \lceil 2/\epsilon \rceil$. Now lay a grid on the plane with integer coordinates and cells of size $c \times c$. Each $(i,j) \in [0,c)^2$ corresponds to a possible offset for the grid. For a given grid position, we eliminate all sensor-mission edges not fully contained within a single cell. Within any cell, there are $O(\Delta c^2)$ sensors and missions; therefore we can find the optimal assignment

---

[2] Although we focus on the plane, it is easy to extend to a fixed higher dimension D.

restricted to that cell by enumerating all $O((\Delta c^2)^{\Delta c^2})$ possible assignments. The solution for a given offset pair (i, j) is the union of the solutions for the individual cells. We compute the solution for each possible offset pair.

We now justify our initial assumption. If the points lie in an extremely large region, then the method as stated may not run in polynomial time, since there may be exponentially many cells to check. This can be easily fixed. First, notice that there will be at most polynomially many non-empty cells, which can be found by iterating through the point coordinates. For each non-empty cell, we can "grow" it outward, to obtain a maximally non-empty region. Performing this action on every non-empty cell (i.e., Union-Find) produces a polynomial collection of independent regions. Now simply run the original algorithm on each independent region, rather than on the entire space.

Proposition 7. *Algorithm 2 is a PTAS.*

Proof. Consider the optimal star-set OPT with total profit P_{opt}. By the Shifting Lemma [11], there must be some vertical offset j that crosses a subset of OPT with total profit at most P_{opt}/c. Similarly, there must be some horizontal offset i that crosses a subset of OPT with total profit at most P_{opt}/c. Therefore the union of the cell-optimal solutions for this (i, j) will be within factor $(1 - 1/c)^2 \geq 1 - 2/c => \geq 1 - \epsilon$ of the optimal. QED

6   Conclusion

In this paper, we introduced a sensor-mission matching problem. We analyzed its complexity, defined constrained versions, and presented approximation algorithms for them. There are many open problems, such as:

- Seek efficient constant-factor approximation algorithms algorithms for *Geom SMD*.
- Seek LP-based $(\Delta + 1)/2$ approximations than run in bounded time.
- Close the approximation gap for Δ-SMD and k-SET PACKING.
- Define online variants of the problem that would admit nontrivial competitive algorithms.

References

1. Ahuja, R., Magnanti, T., Orlin, J.: Network Flows. Prentice-Hall, Englewood Cliffs (1993)
2. Berman, P.: A d/2 approximation for maximum weight independent set in d-claw free graphs. In: Proceedings of SWAT, pp. 214–219 (2000)
3. Byers, J., Nasser, G.: Utility-based decision-making in wireless sensor networks. In: Proceedings of the IEEE Workshop on Mobile and Ad Hoc Networking and Computing (2000)
4. Clark, B., Colbourn, C., Johnson, D.: Unit disk graphs. Discrete Math 86, 165–177 (1990)

5. de Vries, S., Vohra, R.V.: Combinatorial auctions: a survey. INFORMS J. on Computing 15-3, 284–309 (2003)
6. Erlebach, T., Fiala, J.: Independence and coloring problems on intersection graphs of disks (manuscript) (2001)
7. Fleischer, L., Goemans, M.X., Mirrokni, V.S., Sviridenko, M.: Tight approximation algorithms for maximum general assignment problems. In: Proceedings of SODA 2006, pp. 611–620 (2006)
8. Galil, Z., Micali, S., Gabow, H.N.: An O(EV log V) algorithm for finding a maximal weighted matching in general graphs. SIAM J. Comput. 15(1), 120–130 (1986)
9. Hastad, J.: Clique is hard to approximate within $n^{1-\varepsilon}$. Acta Mathematica 182, 105–142 (1999)
10. Hazan, E., Safra, S., Schwartz, O.: On the complexity of approximating k-set packing. Computational Complexity 15(1), 20–39 (2006)
11. Hochbaum, D.S., Maass, W.: Approximation schemes for covering and packing problems in image processing and VLSI. J. ACM 32(1), 130–136 (1985)
12. Hunt III, H.B., Marathe, M.V., Radhakrishnan, V., Ravi, S.S., Rosenkrantz, D.J., Stearns, R.: NC-approximation schemes for NP- and PSPACE-hard problems for geometric graphs. J. Algorithms 26(2), 238–274 (1998)
13. Jain, K., Mahdian, M., Markakis, E., Saberi, A., Vazirani, V.V.: Greedy facility location algorithms analyzed using dual fitting with factor-revealing LP. Journal of the ACM 50(6), 795–824 (2003)
14. Kaplan, L.: Global node selection for localization in a distributed sensor network. IEEE Transactions on Aerospace and Electronic Systems 42(1), 113–135 (2006)
15. Lehmann, D., Mueller, R., Sandholm, T.: The winner determination problem. In: Cramton, Shoham, Steinberg (eds.) Combinatorial Auctions, MIT Press, Cambridge (2006)
16. Marathe, M.V., Radhakrishnan, V., Hunt III, H.B., Ravi, S.S.: Hierarchically specified unit disk graphs. Theor. Comput. Sci. 174(1-2), 23–65 (1997)
17. Matsui, T.: Approximation algorithms for maximum independent set problems and fractional coloring problems on unit disk graphs. In: Proceedings of the Japan Conference on Discrete and Computational Geometry, pp. 194–200 (1998)
18. Mullen, T., Avasarala, V., Hall, D.L.: Customer-driven sensor management. IEEE Intelligent Systems 21(2), 41–49 (2006)
19. Papadimitriou, C.H., Yannakakis, M.: Optimization, approximation, and complexity classes. J. Comput. System Sci. 43, 425–440 (1991)
20. Perillo, M., Heinzelman, W.: Optimal sensor management under energy and reliability constraints. In: Proceedings of the IEEE Conference on Wireless Communications and Networking (2003)
21. Rowaihy, H., Eswaran, S., Johnson, M., Verma, D., Bar-Noy, A., Brown, T., La Porta, T.: A survey of sensor selection schemes in wireless sensor networks. In: SPIE Defense and Security Symposium Conference on Unattended Ground, Sea, and Air Sensor Technologies and Applications IX (2007)
22. Shih, K., Chen, Y., Chiang, C., Liu, B.: A distributed active sensor selection scheme for wireless sensor networks. In: Proceedings of the IEEE Symposium on Computers and Communications (June 2006)
23. Sung, S.C., Vlach, M.: Maximizing weighted number of just-in-time jobs on unrelated parallel machines. J. Scheduling 8-5, 453–460 (2005)
24. Zhao, F., Shin, J., Reich, J.: Information-driven dynamic sensor collaboration. IEEE Signal Processing Magazine 19(2), 61–72 (2002)

Maximal Breach in Wireless Sensor Networks: Geometric Characterization and Algorithms

Anirvan Duttagupta[1], Arijit Bishnu[2], and Indranil Sengupta[2]

[1] Nucleodyne Systems Inc., CA 95014, USA
anirvan@nucleodyne.com
[2] Computer Science and Engg. Dept, Indian Institute of Technology, Kharagpur-721302
{Arijit.Bishnu,Indranil}@iitkgp.ac.in

Abstract. Coverage is a measure of the quality of surveillance offered by a given network of sensors over the field it protects. Geometric characterization of, and optimization problems pertaining to, a specific measure of coverage - maximal breach - form the subject matter of this paper. We prove lower bound results for maximal breach through its geometric characterization. We define a new measure called *average* maximal breach and design an optimal algorithm for it. We also show that a relaxed optimization problem for the proposed measure is NP-Hard.

1 Introduction

Recent advances in wireless technologies coupled with theoretical work have led to centralised and distributed algorithms for various information processing tasks using low cost and low power devices. All these have made Wireless Sensor Networks (*WSN*s) a common and effective solution in a wide range of applications. Research effort in the past few years, in the area of *WSN*s, has become the meeting ground of researchers in signal processing and embedded computation [5], [6], network architecture and protocols [7], distributed algorithms [2] and computational geometry [1], [3], [4], to name just a few.

In this paper, we dwell on one of the basic problems of *WSN*s, viz., *Coverage* [1,2,8]. Coverage is a generic name for a class of measures that quantify the quality of surveillance offered by a given network of sensors over the field it protects. Geometric characterization of, and optimization problems pertaining to, a specific measure of coverage - Maximal Breach - form the subject matter of this paper. Problems related to single-pair maximal breach was first explored in [1]. The coverage problem can be viewed from two angles - the intruder's view and the defender's view. These two view points give rise to two generic combinatorial optimization problems - searching for "safe" paths in the field (important for the intruder) and optimizing the degree of coverage over all parts of the field (important for the defender).

The principle contributions of this paper are the following: (i) mathematical formulation of the problem of optimizing the maximal breach coverage measure for *WSN*s; (ii) a negative lower bound result on single-pair maximal breach; (iii) a simple but important extension to the maximal breach measure - all-pairs average maximal breach - and an optimal algorithm for computing it; (iv) for average maximal breach - a lower-bound result analogous to the single-pair case and an NP-Hardness result.

M. Kutyłowski et al. (Eds.): ALGOSENSORS 2007, LNCS 4837, pp. 126–137, 2008.

In the next section, we present some basic definitions, followed by a brief account of the existing literature on maximal breach. In Section 3, we present lower-bound result for single-pair maximal breach. This leads to Section 4, where we define *All-Pairs Average Maximal Breach*, and deal with its computation and optimization in Section 5. We end with some pointers to future work in Section 6.

2   Background and Prior Work

The bulk of this section is based on [1] and [8]. The *Maximal Breach* measure of *WSN* coverage was first proposed in [1]. To distinguish it from a closely related measure we propose in Section 4, we shall refer to it as *single-pair* maximal breach.

Suppose, $S = \{s_1, s_2, \ldots, s_N\}$ be a set of N sensors deployed over a field modelled as a unit square region A. Each sensor node is a point $s_i = (x_i, y_i) \in A$, where $x_i, y_i \in I\!R$. The *intruder* has complete knowledge of the coordinates of *all* the sensors in S.

2.1   Maximal Breach

Suppose an intruder tries to traverse the field from an initial point i to a final point f. We denote *points* within A with lower-case letters and *paths* with upper-case letters. Consider a path $P(i, f)$ through the field from i to f.

Definition 1 *[Breach] [1]. The quantity* breach *is defined as the* minimum *Euclidean distance from* $P(i, f)$ *to any sensor in the field.*

In A, there are infinitely many paths connecting i and f. One of these has a special property:

Definition 2 *[Maximal Breach Path] [1]. Among the infinitely many paths connecting* i *and* f*, one that has the* maximum *breach value is called a* maximal breach path, $P_b(i, f, S)$*. Maximal Breach,* $breach_{max}(i, f, S)$*, is the breach of the maximal breach path.*

For the intruder, the maximal breach path is the best path to take within the field, because the closest sensor encountered is at the farthest possible distance.

2.2   Prior Work on Maximal Breach

There are uncountably many paths connecting any pair of points in A. How do we find the special path $P_b(i, f, S)$? A fundamental result is established in [1] that reduces the search space (set of candidate paths) to a *finite* size: *At least one maximal breach path must lie along the edges of the* bounded Voronoi diagram *[10] determined by the sensor nodes S and the boundaries of the unit-square field A.*

In the algorithms developed in [1] and [8], a weighted, undirected graph, called the associated graph, G_{VD} is computed from $VD(S)$, where $VD(S)$ denotes the Voronoi diagram for S bounded by A. For each voronoi vertex in $VD(S)$, there is a node in G_{VD}. Additionally, the peripheral edges of $VD(S)$ intersect the boundaries of A. There is a node in G_{VD} for each such intersection point. Finally, the four corners of A are added to the node set of G_{VD}. There is an edge $E = (u, v)$ in G_{VD} iff the corresponding points in $VD(S)$ are connected by a voronoi edge or are adjacent points on the boundary of A. In the former case, the weight $w(E)$ is set to $breach(E)$ (a quantity proportional to the length $|\overline{ss'}|$, where s, s' are the sensors that share the voronoi edge); in the latter case, the weight $w(E)$ is set to its distance from the nearest site. See [1], [8] for details.

Megerian et al. [1] use a combination of *BFS* and binary search on G_{VD} to compute maximal breach. But, they assumed integral Euclidean distances in their algorithm - an unreasonable assumption, given that $x_i, y_i \in I\!R$. In [8], we have published a centralised polynomial time algorithm for maximal breach that gives *exact results* (does not need the integral weights assumption of [1]) and also computes the maximal breach *path* at no extra run-time cost. It uses network flow concepts and computes the maximal breach measure, as well as the *path*, in $O(N \log N)$ time.

3   Single-Pair Maximal Breach - A Lower-Bound Result

We begin this section with some notations and equations.

3.1   Notation

– $VE(S)$, $VV(S)$ denote the set of edges and vertices in $VD(S)$ respectively. $G_{VD} = (V_{VD}, E_{VD})$ denotes the *associated graph* for $VD(S)$.
– $P_b(i, f, S)$ denotes a maximal breach *path* in $\langle A, S \rangle$ and $b^S(i, f)$ the breach *value*.
– $e_{cr}^b(i, f, S)$ denotes the critical edge in $P_b(i, f, S)$, defined further down.

3.2   General Equations for Breach

Let $d(x, y)$ denote the euclidean distance between points x and y. Given S and a point $p \in A$, the closest sensor observability at p [2] is defined as

$$I_C(p, S) = \min_{s \in S} d(s, p). \tag{1}$$

We define breach in terms of $I_C(p, S)$. For a path P in A connecting the points $i, f \in A$, $breach(P)$ is the minimum I_C value over all points on P. Let $\Pi(i, f)$ denote the set of all (infinitely many) paths, within A, connecting i and f. The maximal *breach* between i and f is defined as

$$\mathbf{breach_{max}(i, f, S)} = \max_{P \in \Pi(i,f)} breach(P) = \max_{P \in \Pi(i,f)} \min_{p \in P} \mathbf{I_C(p, S)} \tag{2}$$

and any path that attains this breach value is a maximal breach *path*.

Breach in G_{VD}. We mentioned above that the algorithms in [1] and [8] use the graph G_{VD} for computing maximal breach. In G_{VD}, an edge $e \in E_{VD}$ is assigned a weight $w(e)$ proportional to the distance of E from its nearest sensor, where E is the $VD(S)$ counterpart of e. Let the nodes $s, t \in V_{VD}$ correspond to the points $i, f \in VV(S)$. In G_{VD}, the set $\Pi(s, t)$ of all paths connecting s and t, is finite. For G_{VD}, the problem of computing the maximal breach path between s, t can be expressed succinctly as per the following equation.

$$breach_{max}(s, t, G_{VD}) = \max_{P \in \Pi(s,t)} \min_{e \in P} w(e). \tag{3}$$

There is a special edge, the critical edge e_{cr}^b, in G_{VD} (and correspondingly, in $VD(S)$), which defines the value of maximal breach.

Definition 3 *[Critical Edge]. A critical edge e_{cr}^b is characterized by the following properties: (i) $breach_{max}(i, f, S) = w(e_{cr}^b)$; (ii) e_{cr}^b corresponds to the lightest edge in $P_b(i, f, S)$; (iii) take an arbitrary path $P(i, f)$ connecting i, f along a sequence of voronoi edges. Let $e \in G_{VD}$ be the counterpart of the lightest edge in P. Then $w(e) \leq w(e_{cr}^b)$.*

In short, the maximal breach in $\langle A, S \rangle$ is the weight of the critical edge. See [1] and [8] for details.

3.3   Optimizing Maximal Breach

The coverage optimization problem is, in general terms: *Given a number of sensors, how to deploy them so as to achieve the maximum coverage at every point on the field.* For maximal breach, the optimization problem is a *minimization* problem. The defender would try to secure the weakest segments of the field by reducing the minimum distance from a sensor that the intruder *must* encounter along any path. We look at *two* flavors of optimization, and accordingly, frame two optimization problems.

Problem 1. **[P1]** *(Optimal Coverage)* Given A, two points i and f in A, and a set of N sensors, find an arrangement of the sensors such that $breach_{max}(i, f, S)$ is minimized. Here, the defender optimizes the breach (coverage) value with a fixed number of sensors.

Problem 2. **[P2]** *(Optimal Number of Sensors)* Given A, two points i and f in A, and a positive real number T, find the smallest set of sensors S such that $breach_{max}(i, f, S) \leq T$. Here, the defender tries to meet a breach threshold with a minimum number of sensors.

We give a constructive proof of **[P1], [P2]** having trivial solutions. For this, we need four lemmas describing the behavior of maximal breach under insertion/deletion of sensors into/from S. Their proofs, omitted here for space considerations, can be found in [14].

3.4   Maximal Breach Under Insertion and Deletion of Sensors

The first two lemmas describe the monotonicity of maximal breach, and the latter two describe certain "critical" regions such that addition/removal of sensors to/from those regions affect the maximal breach path.

Lemma 1. *Let S' be an arrangement of sensors over A formed by adding one or more sensors to the configuration S. Then, for any two points i, f $\in$ A, $breach_{max}(i, f, S') \leq breach_{max}(i, f, S)$.*

Lemma 2. *Let S' be an arrangement of sensors over A formed by deleting one or more sensors from the configuration S. Then, for any two voronoi vertices i and f common to both $VD(S)$ and $VD(S')$, $breach_{max}(i, f, S') \geq breach_{max}(i, f, S)$.*[1]

Let $C(p, q, r)$ denote the circle through the points p, q and r.

Lemma 3. [2] *Let a and b be the endpoints of the critical edge e_{cr} of $P_b^S(i, f)$, the maximal breach path in $VD(S)$, and let s_0 and s_1 be the corresponding sites. Then $A^{ins}(i, f, S) = C(s_0, a, s_1) \cup C(s_0, b, s_1)$ is a region such that insertion of any point s within it will guarantee that $b^{S^+} = breach_{max}(i, f, S^+) < breach_{max}(i, f, S) = b^S$, where $S^+ = S \cup \{s\}$.*

Lemma 4. *Let s_0 and s_1 be the sites across the critical edge e_{cr} of $P_b^S(i, f)$, the maximal breach path in $VD(S)$, and let $A^{del}(i, f, S) = \{s_0, s_1\}$. Then the deletion of a point $s \in S$ will guarantee $b^{S^-} = breach_{max}(i, f, S^-) > breach_{max}(i, f, S) = b^S$ if and only if $s \in A^{del}(i, f, S)$, where $S^- = S \setminus \{s\}$.*

3.5 Non-existence of a Lower Bound on Single-Pair Maximal Breach

We now establish that optimizing the *single-pair* maximal breach is a trivial problem.

Theorem 1. *Given a unit-square-field A, two points i and f in A and any positive number B, there exists a set of sensors S_{min}, $|S_{min}| = 8$, such that $B \geq breach_{max}(i, f, S_{min})$.*

Proof. We prove the theorem by constructing the set S_{min}. We start with a set of sensors $S = \{s_1, s_2, \ldots, s_N\}$ chosen as follows. Pick any positive number δ and draw circles, $C_\delta(i)$ and $C_\delta(f)$, of radius δ centred on i and f respectively (δ should be small enough so that $C_\delta(i)$ and $C_\delta(j)$ do not overlap). Place three sensors at random on the circumferences of each of $C_\delta(i)$ and $C_\delta(f)$. Call the subset of S made up solely of these 6 dummy sensors S_d. Place the remaining $N - 6$ sensors at random within $A \setminus [C_\delta(i) \cup C_\delta(f)]$. Note that i are f are forced to be vertices in $VD(S)$.

Now compute $b^S(i, f)$. If $b^S(i, f) \leq B$, well and good. Else, repeatedly augment S, one sensor at a time, in the critical region $A^{ins}(i, f, S)$ until $b^S(i, f) \leq B$. Note that, by virtue of Lemma 3, we are bound to end up with such a set.

Finally, let $e_{cr}(i, f, S)$ be the critical edge of $P_b(i, f, S)$ and s_p, s_q the corresponding sites. Set $S_{min} = \{s_p, s_q\} \cup S_d$. Delete all sensors in $S \setminus S_{min}$. Clearly, by Lemma 4, this does not increase maximal breach. Thus $breach_{max}(i, f, S_{min}) \leq B$ and S_{min} is our required set.[3]   □

Figure 1 provides an intuitive view of the foregoing theorem - maximal breach between the points i, f in the figure can be reduced arbitrarily if we cluster the sensors along the segment $\overline{ab}$ and slide $\overline{ab}$ towards one of the corners. But, in clustering all our sensors on $\overline{ab}$ or $\overline{a'b'}$, we leave a considerably large region $a'b'fa'$ unattended. This is the primary motivation behind the *all-pairs average maximal breach* measure.

---

[1] If $VD(S)$ and $VD(S')$ do not have a common pair of vertices, choose any two corners of A for i and f.

[2] Note that, Lemma 3 expresses a *sufficient* condition for the alteration of maximal breach path, but not a *necessary* one.

[3] We could have made $breach_{max}(i, f, S) = 0$ by placing sensors at i and f. However, consider a situation where it is impossible to place sensors on i or f (e.g., i, f could be located on water bodies). The geometric significance of the above theorem is in the existence of *alternate* locations in A where sensors can be placed to reduce $breach_{max}(i, f)$ without bound.

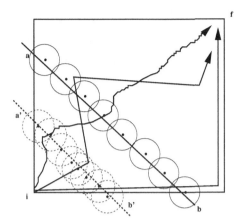

Fig. 1. No lower-bound on breach. The value of $breach_{max}(i, f, S)$ can be reduced arbitrarily.

4 All-Pairs Average Maximal Breach

Until this point, the object of study has been *single-pair* maximal breach. We have established that it is trivial to optimize single-pair maximal breach between any designated pair of points i and f. We now propose and study a new measure - all-pairs *average* maximal breach - a straight-forward extension of its single-pair counterpart.

4.1 Average Maximal Breach

Instead of confining ourselves to a fixed pair of starting and ending coordinates (of the intruder), we consider all possible pairs of points (i, f) within A. We determine the critical edge of the maximal breach path between each pair of points, and then take the average. We make our domain finite by restricting the set of feasible starting and ending positions to the set of vertices of $VD(S)$. This makes sense, since a voronoi vertex is the center of a maximum empty circle [9], [10]. It is precisely the point that has the maximum value of $I_C(p, S)$ (Equation 1) in its immediate vicinity. So, an intruder would always prefer to land up on a voronoi vertex.

Definition 4 *[Average Maximal Breach].* Let $E_b(S) = \{e \in E_{VD} \mid \exists i, j \in V_{VD}$ such that $e_{cr}^b(i, j, S) = e\}$. In other words $E_b(S)$ is the subset of E_{VD} made up of only the critical edges of maximal breach paths in $\langle A, S \rangle$. Then,

$$avgBreach_{max}(S) = \frac{\sum\limits_{e \in E_b(S)} w(e)}{|E_b(S)|}. \tag{4}$$

Clearly, $|E_b|$ is $O(N)$. Lemma 5 proves that $E_b(S)$ is nothing but the *Maximum Cost Spanning Tree* [13] of G_{VD}. Thus $|E_b| = |V_{VD}| - 1$.

4.2 Significance of the Measure

The average breach measure is designed to meet the following goals:

1. An *optimal* value of $avgBreach_{max}$ provides the desired level of coverage *uniformly* all over the field because the set of vertices of G_{VD} (the voronoi vertices and the intersection of the voronoi edges with the boundaries of A), is uniformly dispersed over A.
2. The measure should be *sound* - it should conform to the intuitive requirement that adding sensors to S gives better coverage.
3. The measure should be useful in practical scenarios. This is true for $avgBreach_{max}$ because we take into account all *reasonable* trajectories of the intruder within A.

5   An Optimal Algorithm for Computing Average Maximal Breach

In [8], we have developed a greedy algorithm that computes the single-pair maximal breach path. In this section, we develop an *optimal*, greedy algorithm for computing *all-pairs average* maximal breach. The algorithm hinges upon the following lemma.

5.1   Maximal Breach Path and *Maximum* Cost Spanning Tree

The *Maximum Cost Spanning Tree (MaxST)* of a connected graph can be computed by an $O(|E| \log |E|)$ greedy algorithm that parallels Kruskal's Minimum Cost Spanning Tree algorithm [13]. The associated graph G_{VD} is connected, and $|E_{VD}| = O(N)$. Thus, $MaxST(G_{VD})$ can be computed in $O(N \log N)$ time by first sorting E_{VD} in *descending* order of weights and then running Kruskal's algorithm on G_{VD}. The only difference is that at each step we pick the *heaviest* feasible edge, instead of the lightest one.

Lemma 5. *Suppose T is a Maximum Cost Spanning Tree of G_{VD} computed by the greedy algorithm outlined above. Pick any pair of nodes $s, t \in V_{VD}$. Then the path $P_T(s, t)$ in T between s and t is also a maximal breach path between s and t in G_{VD}.*

Proof. Call the edges in T the *branch* edges, denoted by b_i, and the edges in $G_{VD} \setminus T$ the *arc* edges, denoted by a_j. Let $E_T = \{b_1, b_2, \ldots, b_{|V_{VD}|-1}\}$ denote the set of all branches and $P_T(s, t) = \langle b_{l_1}, b_{l_2}, \ldots, b_{l_k} \rangle$. Also, let $P'(s, t) = \langle e_{r_1}, e_{r_2}, \ldots, e_{r_i}, \ldots, e_{r_p} \rangle$ be another path, in G_{VD}, between s and t.

Suppose $e_{r_i} = a_j = (u, v)$ is the first *arc* in the sequence $P'(s, t)$. Then, $b_{l_s} = e_{r_s}$, $1 \le s \le i - 1$ (because $P_T(s, t)$ is a *unique* path in T). Now, a_j could have been omitted from T for two reasons:

1. All the branches were considered before a_j. Then, $w(a_j) \le \min\{b \mid b \in E_T\}$. In this case, the minimum edge in $P_T(s, t)$ is heavier than that of P'.
2. The introduction of a_j would have created a cycle $s \rightsquigarrow u \rightarrow v \rightsquigarrow s$ in the spanning forest of G_{VD}. This implies, at the point of time a_j was considered, there already existed a path $P_T(s, v)$ between s and v, using *only* branches encountered *before* a_j.

Since we wish to maximise the minimum edge of P', we can discard the prefix $\langle e_{r_1}, \ldots, a_j \rangle$ in favour of $P_T(s, v)$.

In this manner, all arcs a_j can be eliminated. They either do not figure as the critical edge, or can be discarded in favour of alternate paths comprising only branches.   □

Lemma 5 leads directly to a simple algorithm for computing all-pairs average breach. The algorithm follows.

5.2   The Algorithm

Input: $G_{VD} = (V_{VD}, E_{VD})$, the associated graph of $VD(S)$.
Output: $avgBreach_{max}(S)$.
Method: See Algorithm 1.

Algorithm 1. MaxSTAverageBreach

1: **Variables:**
2:   F: *Set* of *Set*s and E: *Array* of *Edge*s. {F stores the spanning forest of G_{VD} at all times.}
3:   e: *Edge*; u, v: *Node*.
4:   T_u, T_v, T: *Set* of *Edge*s.
5:   B_{max}: $|V_{VD}| \times |V_{VD}|$ two-dimensional *Array* of *Edge*s. {$B_{max}[i, j]$ stores the critical edge of the maximal breach path between i and j.}
6: $F \leftarrow \{\{1\}, \{2\}, \ldots, \{|V_{VD}|\}\}$. {The nodes of G_{VD} are numbered $1, 2, \ldots$}
7: $E \leftarrow E_{VD}$.
8: Sort E in *descending* order of weights.
9: **while** $|F| > 1$ **do**
10:    $e \leftarrow E.pop\_first()$.
11:    $u \leftarrow e.source(), v \leftarrow e.target()$.
12:    $T_u \leftarrow Find(u), T_v \leftarrow Find(v)$.
13:    **if** $T_u \neq T_v$ **then**
14:       $F.remove(T_u), F.remove(T_v)$.
15:       $\forall i \in T_u, \forall j \in T_v, B_{max}[i, j] \leftarrow e$.
16:       $T \leftarrow Union(T_u, T_v)$.
17:       $F.insert(T)$.
18:    **end if**
19: **end while**
20: $avgBreach \leftarrow \frac{COST(F)}{|V_{VD}| - 1}$.

5.3   Correctness and Analysis

Algorithm 1 makes just one addition to Kruskal's algorithm: the assignment in line 15, where the critical edge between a set of node-pairs (i, j) is actually determined. The following lemma justifies the assignment.

Lemma 6. *Let $F = \{T_1, T_2, \ldots, T_n\}$ be the Maximum Cost Spanning Forest of G_{VD} just before an edge $e = (u, v)$ is added to F. Let T_u and T_v be trees in F to which the endpoints u and v of e belong. Then, $\forall i \in T_u, \forall j \in T_v$, the critical edge between i and j is e.*

Proof. Consider nodes $i \in T_u$ and $j \in T_v$. See Fig. 2. Adding $e = (u, v)$ to F connects T_u and T_v and introduces a path $P_T(i, j)$ between i and j. Moreover, $P_T(i, j)$ is the path connecting i and j in the Maximum Cost Spanning Tree constructed by the algorithm. Thus, by Lemma 5, $P_T(i, j)$ is a maximal breach path between i and j, and e is one of its edges. But, since the algorithm picks heavier edges first, $w(e) \leq \min\{b \mid b \in T_u \cup T_v\}$. Thus, e must be the critical edge of $P_T(i, j)$.   □

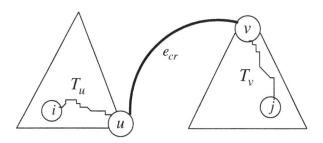

Fig. 2. Computing critical edges bottom-up

Lemma 6 helps us prove the following loop-invariant for Algorithm 1.

Lemma 7. *Let $F = \{T_1, T_2, \ldots, T_n\}$ be the Maximum Cost Spanning Forest of G_{VD} at the end of the ith iteration $(1 \leq i \leq |V_{VD}| - 1)$ of the loop starting at line 9. Then, for all nodes x and y that are connected in F, the critical edge of the maximal breach path between x and y is known, and does not change thereafter.*

Proof. Let b_i denote the branch added to F during the ith iteration, $1 \leq i \leq |V_{VD}| - 1$. The proof is by induction on i.

For the base case $(i = 1)$, note that $b_1 = (u_1, v_1)$ is the heaviest edge in G_{VD}. So b_1 constitutes the maximal breach path, as well as the critical edge, between u_1 and v_1. Also u_1 and v_1 are the only nodes connected in F at this point. Thus the loop invariant holds at the end of iteration 1.

Suppose the invariant holds at the end of some iteration $i-1$. During the ith iteration, the new branch $b_i = (u_i, v_i)$ joined the trees T_{u_i} and T_{v_i}, and connected exactly $|T_{u_i}| \times |T_{v_i}|$ new node-pairs in F. By Lemma 6, b_i is the critical edge for all these pairs. Hence, the invariant holds after the ith iteration as well.   □

Finally, we have the following theorem.

Theorem 2. *Algorithm 1 is optimal and computes the the average breach over all pairs of nodes in G_{VD} in $O(N^2)$ time.*

Proof. Firstly, the algorithm halts because at each iteration exactly one edge is added to F, until there is a single connected component in F. This outcome is guaranteed because G_{VD}, by definition, is connected.

When the algorithm terminates, all nodes in G_{VD} are connected by F. Thus, by Lemma 7, the critical edges between all pairs of nodes is correctly known.

For computing the time-complexity, recall that $|E_{VD}| = O(N)$. Lines 6 and 7 take $O(N)$ time. The sorting in line 8 takes $O(N \log N)$. Within the while loop of line 9, the operations of lines 10 and 11 take constant-time. The *Union-Find* operations (lines 12 and 16) can be done in $O(\log^* N)$. And the set operations of lines 14 and 17 can be done in linear time. Since the while loop runs $O(N)$ times, in the absence of line 15, the loop-complexity would have been $O(N^2)$.

The costliest operation in the loop is done in line 15. By aggregate analysis, we need to populate $O(N^2)$ entries. Thus the entire algorithm runs in $O(N^2)$ time.

The algorithm is optimal because it computes $O(N^2)$ quantities in $O(N^2)$ time.   □

5.4   Non-existence of a Lower Bound on Average Maximal Breach

We have, as a counterpart of Theorem 1, a negative result.

Theorem 3. *Let $F \subset A$ be a set of N points in A. The points in F act as feasible starting and ending points for an intruder dropped inside A. For any choice of F, and any positive real number B, there exists a set S of $O(N^2)$ sensors such that $avgBreach_{max}(S) \le B$.*

Proof. Pick any pair of points i and f from F. By Theorem 1, there exists a set of sensors $S_{i,f}$, $|S_{i,f}| \le 8$, such that $breach_{max}(i, f, S_{i,f}) \le B$. Now, this is true for *all* $i, f \in F$. Observe that, as far as satisfying the breach upper-bound B is concerned, each pair of points i, f can be treated *independently*. This is because additional sensors can only decrease the value of $breach_{max}(i, f, S_{i,f})$, by Lemma 1. So, let

$$S = \bigcup_{i,f \in F} S_{i,f}.$$

It follows from the preceding argument that $B \ge breach_{max}(i, f, S)$, $\forall i, f \in F$. Hence, $B \ge avgBreach_{max}(S)$. Moreover, since we have $O(N^2)$ pairs of points, and 8 sensors for each pair, $|S| = O(N^2)$.   □

Like in the case of single-pair maximal breach, the theorem above says that all-pairs average breach has no lower bound for any given set of points acting as feasible starting and ending positions of an intruder's tours through A. However, there is one fundamental difference. In case of single-pair breach, any breach threshold can be met with a *constant* number of sensors. For all-pairs average breach, however, we could potentially need $O(|F|^2)$ sensors. As we shall show below, the problem of minimizing the number of sensors while meeting a given average breach threshold is a non-trivial (in all probability, NP-Hard) problem.

5.5   Optimization/Decision Problems in Terms of Average Measures

Theorem 3 states that with at most $8|F|^2$ sensors (where F is the set of feasible starting and ending points for the intruder), any average breach threshold is achievable. But there is no guarantee that this is the *optimal* number of sensors. Next, we concentrate on the problem of meeting the threshold with the *optimal* number of sensors. The average breach version of problem **[P2]** can be framed in a manner identical to Section 3.3, with the measure $avgBreach_{max}$ substituting for $breach_{max}(i, f)$.

5.5.1 Decision Problem Restricted to Finite Domain

Note that the solution space of the problem is uncountable, and as such, not combinatorial in nature. A has uncountably many feasible positions at which sensors from S might be placed. We shall follow the technique used by [11] to restrict the feasible solution space to a *finite* size.

We restrict the field A to a set of N discrete points on the plane. Without loss of generality as in [11] we can restrict the points in A to ones with integral coordinates (i, j). The set A represents the *feasible* positions for placing sensors, as well as feasible starting and ending positions of tours made by the intruder. In this restricted setup, the decision problem takes the following form.

Problem 3. **[P2-AVG-DEC-FINITE]** Given A, a positive real number T_b and a positive integer n, does there exist a set of points S, $|S| \leq n$, such that $avgBreach_{max}(S) \leq T_b$? We can encode an instance of this problem by the tuple $\langle A, n, T_b(\text{ or } T_s) \rangle$. Clearly, the size of the problem is determined by $|A| = N$.

To the best of our knowledge, the question of hardness of **[P2-AVG-DEC-FINITE]** is open. But we have a result about a "relaxed" version of the problem, stated below as Problem 4.

5.5.2 Maximum-Breach-Finite: An NP-Hard Problem

Problem 4. **[Maximum-Breach-Finite]** Given A, a positive real number T_b and a positive integer n, does a set of points $S \subset A$ exist such that $|S| \leq n$ and for any $i, j \in A$ and *any* path $P(i, j)$ between them, $breach(P(i, j)) \leq T_b$?

In geometric terms, this problem requires us to find a set of points S such that all points in any arbitrary path in A are within a distance T_b from at least one point in S.

Theorem 4. *Maximum-Breach-Finite is* **NP-***Hard*.

Proof. We prove this theorem by reducing **Minimum-Geometric-Disc-Cover** [12] to **Maximum-Breach-Finite**.

An instance I_{mgdc} of **Minimum-Geometric-Disc-Cover** (**MGDC**) is given by $\langle A, T_b, n \rangle$, where the goal is to determine whether the points in A can be covered by at most n discs of radius T_b. The corresponding instance I_{mbf} of **Maximum-Breach-Finite** (**MBF**) is also $\langle A, T_b, n \rangle$, where the interpretation is as given in the theorem statement.

Suppose answer(I_{mgdc}) = **yes**. Then we have a set $\mathbb{C}$ of at most n discs of radius T_b such that A is covered by $\mathbb{C}$. Then, let $S = \{s_i \mid s_i \text{ is the center of the } i\text{th disc in } \mathbb{C}, 1 \leq i \leq |\mathbb{C}|\}$. By hypothesis, for all points $p \in A$, there is an $s \in S$ such that $d(s, p) \leq T_b$. This is sufficient to ensure that $breach(P(i, j)) \leq T_b$ for any points $i, j \in A$ and any path $P(i, f)$ connecting them. Thus answer(I_{mbf}) = **yes**.

Similarly, it can be proved that if answer(I_{mbf}) = **yes**, i.e. there exists a set S, $|S| \leq n$ that satisfies the maximum breach criterion, then $|S|$ disks of radius T_b centered on the points in S will cover A.   □

6   Conclusions and Future Work

Our study of the geometric and combinatorial properties of single-pair and average maximal breach has led to exact polynomial time algorithms for *computing* the measures. We have framed and solved the problem of *optimizing single-pair* maximal breach. For *average* maximal breach, we have proved a "relaxed" problem **NP-Hard**. We have also presented important *lower-bound* results for both the measures. However, we have not been able to decide the complexity of **[P2-AVG-DEC-FINITE]**. We need to address this and solve the problem exactly or approximately, as the case may be.

References

1. Megerian, S., Koushanfar, F., Potkonjak, M., Srivastava, M.B.: Worst and Best-Case Coverage in Sensor Networks. IEEE Trans. on Mobile Computing 4(1), 84–92 (2005)
2. Li, X.-Y., Wan, P.-J.: Coverage in Wireless Ad-hoc Sensor Networks. IEEE Trans. on Computers 52(6), 53–763 (2003)
3. Huang, C.-F., Tseng, Y.-C.: The Coverage Problem in Wireless Sensor Networks. In: Proc. of the 2nd ACM International Conference on Wireless Sensor Networks and Applications, ACM Press, New York (2003)
4. Urrutia, J.: Routing with guaranteed delivery in geometric and wireless networks. In: Handbook of Wireless Networks and Mobile Computing, pp. 393–406. John Wiley & Sons, Chichester (2002)
5. Pottie, G.J., Kaiser, W.J.: Wireless Integrated Network Sensors. Communications of the ACM 43, 51–58 (2000)
6. Borriello, G., Want, R.: Embedded Computation Meets the World Wide Web. Communications of the ACM 43 (May 2000)
7. Ye, W., Heidemann, J., Estrin, D.: An Energy Efficient MAC Protocol for Wireless Sensor Networks. In: Proc. of Infocom 2002, vol. 3, pp. 3–12. IEEE CS Press, Los Alamitos (2002)
8. DuttaGupta, A., Bishnu, A., Sengupta, I.: Optimisation Problems Based on the Maximal Breach Path Measure for Wireless Sensor Network Coverage. In: Madria, S.K., Claypool, K.T., Kannan, R., Uppuluri, P., Gore, M.M. (eds.) ICDCIT 2006. LNCS, vol. 4317, Springer, Heidelberg (2006)
9. de Berg, M., Van Kreveld, M., Overmars, M., Schwarzkopf, O.: Computational Geometry Algorithms and Applications. Springer, Heidelberg (1997)
10. Aurenhammer, F.: Voronoi Diagrams - A Survey of a Fundamental Geometric Data Structure. ACM Computing Surveys 23(3) (September 1991)
11. Fowler, R., Paterson, M., Tanimoto, S.: Optimal Packing and Covering in the Plane are NP-Complete. Information Processing Letters 12(3), 133–137 (1981)
12. Hochbaum, D.S., Maass, W.: Approximation schemes for covering and packing problems in image processing and VLSI. JACM 32, 130–136 (1985)
13. Papadimitriou, C.H., Steiglitz, K.: Combinatorial Optimization: Algorithms and Complexity, ch. 12. Prentice Hall, Englewood Cliffs (2001)
14. DuttaGupta, A., Bishnu, A., Sengupta, I.: Maximal Breach in Wireless Sensor Networks: Geometric Characterization and Algorithms. In: ALGOSENSORS 2007. Preproceedings, pp. 142-163 (2007) ISBN 978-83-7125-158-0

Counting-Sort and Routing in a Single Hop Radio Network⋆

Maciej Gębala and Marcin Kik

Institute of Mathematics and Computer Science,
Wrocław University of Technology
Wybrzeże Wyspiańskiego 27, 50-370 Wrocław, Poland
Maciej.Gebala@pwr.wroc.pl, Marcin.Kik@pwr.wroc.pl

Abstract. We consider two problems. First, sorting of n integer keys from the $[0, 2^m - 1]$ range, stored in p stations of a single-hop and single channel radio network. Second problem is routing of the packets between the stations of the network. We introduce counting-sort algorithm which has $3mr_i + s_i + d_i + 3$ energetic cost and $nm + n + p$ time cost, where station a_i stores s_i keys (r_i distinct keys) and receives d_i keys. On the basis of this sorting, we construct routing protocols with energetic costs $(3\lceil \log_2 p \rceil + 2)r_i + s_i + d_i + 5$ and $(3\lceil \log_2 p \rceil + 4)r_i + s_i + d_i + 6$, and time costs $n\lceil \log_2 p \rceil + n + 3p$ and $r\lceil \log_2 p \rceil + n + r + 3p$, respectively, where r is sum of all r_i. Our routing is attractive alternative for previous solutions, since it is efficient, deterministic and simple.

1 Introduction

Radio network is a distributed system with no central arbiter, consisting of p radio transceivers called *stations*. Stations are usually small devices running on batteries. Therefore, it is of big importance to design protocols for radio networks with power efficiency in mind, i.e., the station must consume as little power as possible. We assume that each of p stations belonging to the radio network has unique ID – an integer in the $[0, p - 1]$ range. We consider only static radio networks where the number of station is fixed and no members join or leave the network during the protocol operation. In this paper we focus on single-hop radio networks where each single station lies within the transmission range of all other stations. Finally, we only consider model with single channel of communication. A station uses energy only when its transceiver is active, i.e. while sending or receiving any information. When transceiver is inactive the energy consumption is very small and therefore is ignored. We assume that a station uses one unit of energy for sending or receiving single message.

Let p denote the number of stations. For $0 \leq i \leq p - 1$, station a_i initially stores s_i items (with r_i distinct values) and is destination of d_i items from q_i other stations. By $n = \sum_{i=0}^{p-1} s_i$ we denote total number of items. Let $r = \sum_{i=0}^{p-1} r_i$ and $q = \sum_{i=0}^{p-1} q_i$. We assume that during a single round each station can send

---

⋆ Partially supported by KBN grant 3 T11C 011 26.

M. Kutyłowski et al. (Eds.): ALGOSENSORS 2007, LNCS 4837, pp. 138–149, 2008.

or receive no more than single message containing either single key or an integer
between 0 and n.

We consider two problems. First, sorting n integer keys of range $[0, 2^m - 1]$
stored in p stations of a single-hop and single channel radio network. Second,
problem of routing of the packets between the stations of the network.

1.1   Previous Works

Some energy efficient sorting algorithms are described in [7,8,2,3]. Singh and
Prasanna [7,8] proposed sorting algorithm based on quick-sort and balanced
selection with $\Theta(\log n)$ energy cost and $\Theta(\frac{n}{c} \log n)$ time cost, where c is the
number of communication channels. Sorting based on merging and an algorithm
for merging with energetic cost $O(\log^* n)$ has been proposed in [2]. In these
algorithms it is assumed that each station stores single key (i.e. $p = n$). The
algorithm described in [3] extends results from [2] for sorting n keys stored in p
stations (where each station stores $\frac{n}{p}$ keys) with $8\frac{n}{p} \log_2 p + 2(\log_2 p + 1)\log_2 p$
energetic cost and $(3n + 2p - 2)\log_2 p$ time cost.

Some energy efficient permutation routing protocols are described in [5,1]. The
protocol described by Nakano, Olariu and Zomaya routes n packets between p
station (each station stores $\frac{n}{p}$ and is destination for $\frac{n}{p}$ packets) with $(4d+7b-1)\frac{n}{p}$
energetic cost and $(2d+2b+1)\frac{n}{c}+c$ time cost, where $d = \left\lceil \frac{\log \frac{n}{c}}{\log \frac{n}{p}} \right\rceil$ and $b = \left\lceil \frac{\log c}{\log \frac{n}{p}} \right\rceil$.
In such case, if we consider only single channel radio network the cost reaches
$(4 \left\lceil \frac{\log p}{\log \frac{n}{p}} \right\rceil - 1)\frac{n}{p}$ for energy and $(2 \left\lceil \frac{\log p}{\log \frac{n}{p}} \right\rceil + 1)n + 1$ for time. Datta and Zomaya
in [1] presented algorithm with $6\frac{n}{p} + 2p + 8$ energetic cost and $2n + p^2 + p + 2$
time cost. Both algorithms are effective when $p \approx \sqrt{n}$.

In [6] Nakano, Olariu and Zomaya introduced randomized routing protocol
where for every $f \geq 1$ the task of routing n items in p stations can be completed
with probability exceeding $1 - 1/f$ in time $n + O(q + \ln f)$ with energetic cost
below $s_i + d_i + O(q_i + r_i \log p + \log f)$.

1.2   Our Results

In this paper we present two following results (we use the notation introduced
so far)

Theorem 1. *For the single hop and single channel radio network with p stations
there exists sorting algorithm for n integer keys of range $[0, 2^m - 1]$ that works
in $mn + n + p$ rounds of time and where each station a_i uses no more than
$3mr_i + d_i + s_i + 3$ energy.*

Theorem 2. *For the single hop and single channel radio network with p stations
there exist routing algorithms that work*

1. *in $n\lceil \log_2 p \rceil + n + 3p$ rounds of time with each station a_i using no more than
   $(3\lceil \log_2 p \rceil + 2)r_i + s_i + d_i + 5$ energy;*

2. *in* $r\lceil \log_2 p \rceil + n + r + 3p$ *rounds of time with each station* a_i *using no more than* $(3\lceil \log_2 p \rceil + 4)r_i + s_i + d_i + 6$ *energy.*

The second algorithm is much faster than the first one, if $n \gg r$ (and r is always bounded by $\min\{n, p(p-1)\}$).

For p nearing n (for example when $n \approx p\log_2 p$) our routing is more efficient than algorithms described in [5,1]. Besides it is more universal (arbitrary vs. permutation routing). Also comparing with randomized algorithm described in [6] our protocol is simpler and has comparable energy complexity, without randomization.

2   Preliminaries

We assume that each key is an integer in the range $[0, 2^m - 1]$. Let $b_i(key)$ denote ith bit in the binary representation of key (i.e. $key = \sum_{i=0}^{m-1} 2^i \cdot b_i(key)$). Let $g_l(key) = \sum_{i=l}^{m-1} 2^{i-l} \cdot b_i(key)$. For each (*level*) l, $m \geq l \geq 0$, we say that key is in the *group* $G\left(g_l(key), l\right)$ (i.e. $G(g,l) = \{k : 0 \leq k \leq 2^m - 1 \wedge g_l(k) = g\}$). There are 2^{m-l} disjoint groups on level l, partitioning the set $\{0, \ldots, 2^m - 1\}$ into blocks of size 2^l.

The number of the stations in the network is denoted by p. For $0 \leq i \leq p-1$, a_i denotes the ith station of the network. Each station initially stores s_i keys in its local (sorted) table $key[a_i][0 \ldots s_i - 1]$. Let $n = \sum_{i=0}^{p-1} s_i$ denote the total number of keys. Let $POS = \{(i,j) : 0 \leq i < p \wedge 0 \leq j < s_i\}$ (set of positions of elements). And let r_i denote the number of distinct values of the keys in $key[a_i]$. Each a_i stores these values in $key'[a_i][0 \ldots r_i - 1]$. For $0 \leq j \leq r_i - 1$, $c_{i,j}$ denotes the number of copies of $key'[a_i][j]$ in $key[a_i]$. Thus $s_i = \sum_{j=0}^{r_i-1} c_{i,j}$. Let $r = \sum_{i=0}^{p-1} r_i$.

Let d_i be the number of keys for which a_i is final destination, and let q_i be the number of stations that initially stored such keys ($q_i \leq d_i$).

3   Counting-Rank

We use following additional local variables in station a_i:

-   $lrm[a_i]$ – copy of last received message (if needed),
-   $rig[a_i][j]$ – rank of $key[a_i][j]$ in its "current" group,
-   $rng[a_i][j]$ – rank of $key[a_i][j]$ in its "next" group,
-   $gs[a_i][j]$ – number of keys in group of $key[a_i][j]$,
-   $bg[a_i][j]$ – number of keys in groups preceding group of $key[a_i][j]$.
-   $rank[a_i][j]$ – "current" rank of $key[a_i][j]$.
-   $first[a_i][j]$, $last[a_i][j]$ – additional Boolean variables used in routing procedures

We say that "(i,j) is in $G(g,l)$" if and only if $key[a_i][j] \in G(g,l)$.

In the procedure Init, each station a_i learns the total number of keys n and the initial ranks of its keys in the (*single*) group on level m. Note that this initial ranking depends only on the initial positions of the keys and totally ignores their values: $rank[a_i][j] < rank[a_{i'}][j']$ if and only if (i, j) is less than (i', j') in lexicographical ordering.

procedure Init($\langle a_0, \ldots, a_{p-1} \rangle, m$)
begin
    Each station a_i does, for each j: $bg[a_i][j] \leftarrow 0$;
    station a_0 does: **begin**
        $lrm[a_0] \leftarrow 0$;
        foreach $j \in \{0, \ldots, s_0 - 1\}$ **do** $rig[a_0][j] \leftarrow j$;
    end
    for $i \leftarrow 0$ **to** $p - 2$ **do**
        a_i sends $\langle x \rangle$, where $x = lrm[a_i] + s_i$;
        a_{i+1} receives $\langle x \rangle$ and does: **begin**
            $lrm[a_{i+1}] \leftarrow x$;
            foreach $j \in \{0, \ldots, s_{i+1} - 1\}$ **do** $rig[a_{i+1}][j] \leftarrow j + x$;
        end

    In the last time slot: **begin**
        station a_{p-1} broadcasts $\langle x \rangle$, where $x = lrm[a_{p-1}] + s_{p-1}$;
        each other station receives $\langle x \rangle$;
        each station a_i does, for each j: $n[a_i] \leftarrow gs[a_i][j] \leftarrow x$;
    end
    each station a_i does, for each j: **begin**
        if $s_i > 0$ and $rig[a_i][0] = 0$ **then** $rng[a_i][0] \leftarrow 0$;
        $rank[a_i][j] \leftarrow bg[a_i][j] + rig[a_i][j]$;
    end
end

Algorithm 1. Procedure Init

Init is used in the procedure Counting-rank. In the procedure Counting-rank we compute the final ranks of the keys in the sorted sequence of n keys. For equal keys their ordering is the same as in the initial sequence. Hence, our procedure is suitable for *stable* sorting. For each station a_s and for $0 \leq i < n$, we define set of indexes $S(a_s, i)$ as follows: $S(a_s, t) = \{j : bg[a_s][j] < t < bg[a_s][j] + gs[a_s][j]\}$. Note that the value $S(a_s, t)$ depends only on the local variables of a_s and may be computed by internal computations of a_s. Intuitively, it denotes the set positions (s, j) located in a_s that are in the same "current" group as the position with rank t. (We will show in Lemma 2(5), that all positions have consistent view of their current group.)

For $m \geq l \geq 0$, for $(i, j) \in POS$, we say that (i, j) is *classified on level* l if and only if the following conditions are satisfied:

1. $gs[a_i][j]$ is the number of positions (i', j') in $G(g_l(key[a_i][j]), l)$, and
2. $bg[a_i][j]$ is the total number of positions (i', j') in the groups $G(g, l)$ such that $0 \leq g < g_l(key[a_i][j])$, and

procedure Counting-rank($\langle a_0, \ldots, a_{p-1} \rangle, m$)
begin
    Init($\langle a_0, \ldots, a_{p-1} \rangle, m$);
    (* REGROUPING PHASE *)
    for $l \leftarrow m - 1$ **downto** 0 **do**
        for $t \leftarrow 0$ **to** $n - 1$ **do**
            For the (unique) pair (a_{snd}, j') such that $rank[a_{snd}][j'] = t$, a_{snd} does:
            begin
                if $rig[a_{snd}][j'] = 0$ **then** $lrm[a_{snd}] \leftarrow 0$;
                Let $x = lrm[a_{snd}] + (1 - b_l(key[a_{snd}][j']))$;
                if $|S(a_{snd}, t)| < gs[a_{snd}][j']$ and
                ($rig[a_{snd}][j'] = gs[a_{snd}][j'] - 1$ or $j' = s_{snd} - 1$ or
                $bg[a_{snd}][j' + 1] \neq bg[a_{snd}][j']$) **then**
                    a_{snd} sends message $\langle x \rangle$
            end
            For the (at most one) pair (a_{rcv}, j) such that $rank[a_{rcv}][j] = t + 1$ and
            $bg[a_{rcv}][j] \leq t < bg[a_{rcv}][j] + gs[a_{rcv}][j]$, a_{rcv} listens to $\langle x \rangle$ (unless
            $a_{rcv} = a_{snd}$) and does:
            (* CASE A: $key[a_{rcv}][j]$ is successor of $key[a_{snd}][j']$ in its group *)
            begin
                if $b_l(key[a_{rcv}][j]) = 0$ **then** $rng[a_{rcv}][j] \leftarrow x$;
                else $rng[a_{rcv}][j] \leftarrow rig[a_{rcv}][j] - x$;
                $lrm[a_{rcv}] \leftarrow x$;
            end
            Each a_{rcv} such that $\exists_j\ bg[a_{rcv}][j] \leq t = bg[a_{rcv}][j] + gs[a_{rcv}][j] - 1$,
            listens to $\langle x \rangle$ (unless $a_{rcv} = a_{snd}$) and does, for each $j \in S(a_{rcv}, i)$:
            (* CASE B: $key[a_{snd}][j']$ is the last one in its group *)
            begin
                if $b_l(key[a_{rcv}][j]) = 0$ **then**
                    $gs[a_{rcv}][j] \leftarrow x$;
                else
                  $bg[a_{rcv}][j] \leftarrow bg[a_{rcv}][j] + x$;
                  $gs[a_{rcv}][j] \leftarrow gs[a_{rcv}][j] - x$;
                $rig[a_{rcv}][j] \leftarrow rng[a_{rcv}][j]$;
                $rank[a_{rcv}][j] \leftarrow bg[a_{rcv}][j] + rig[a_{rcv}][j]$;
            end
    each a_i, for each j, does: **begin**
        if $rig[a_i][j] = 0$ **then** $first[a_i][j] \leftarrow true$; **else** $first[a_i][j] \leftarrow false$;
        if $rig[a_i][j] = gs[a_i][j] - 1$ **then** $last[a_i][j] \leftarrow true$;
        else $last[a_i][j] \leftarrow false$;
    end
end

Algorithm 2. Procedure Counting-rank

3. $rig[a_i][j]$ is the rank (in the lexicographical ordering by (i, j)) of (i, j) in $G(g_l(key[a_i][j]), l)$, and
4. $rank[a_i][j] = bg[a_i][j] + rig[a_i][j]$ (final result, if we ignore the bits $l - 1, \ldots, 0$ in the keys).

For $m > l' \geq 0$ and $0 \leq t' < n$, let $slot(l', t')$ denote the time slot of RE-GROUPING PHASE in which the variables l and t have values l' and t', respectively. Let $slot(-1, 0)$ denote the first time slot *after* the REGROUPING PHASE. Let $next(t, l)$ denote the next slot after $slot(t, l)$. For $m > l \geq 0$, $next(l, t) = slot(l, t+1)$ if $0 \leq t < n - 1$, and $next(l, n - 1) = slot(l - 1, 0)$.

Lemma 1. *In* Counting-rank*:*

1. *For any $(i, j) \in POS$, if, after some time slot, $rig[a_i][j] = 0$, then in all the following time slots $rng[a_i][j] = rig[a_i][j] = 0$.*
2. *For any $(i, j) \in POS$, before each slot after* Init*, we have $rank[a_i][j] = bg[a_i][j] + rig[a_i][j]$.*

Proof. The code ensures that $rng[a_i][j]$ becomes zero whenever $rig[a_i][j]$ becomes zero. If at the beginning of time slot $rig[a_i][j] = 0$ then $rank[a_j][j] = bg[a_j]$. Thus if $rank[a_i][j] = t + 1$ then $t < bg[a_i][j]$ and a_i will not execute code of CASE A of REGROUPING PHASE (the only fragment that could change $rng[a_i][a_j]$). Consequently $rng[a_i][j]$ and $rig[a_i][j]$ will remain equal to zero. The property 2 can be easily seen from the code.   □

Lemma 2. *For $m - 1 \geq l \geq 0$ and $0 \leq t < n$ or $(l, t) = (-1, 0)$, for each $(i, j) \in POS$, at the beginning of $slot(l, t)$:*

1. *either:*
   - $t < bg[a_i][j] + gs[i][j]$ *and (i, j) is classified on level $l + 1$, or*
   - $t \geq bg[a_i][j] + gs[i][j]$ *and (i, j) is classified on level l,*
   and
2. *if $t \geq rank[a_i][j]$ then $rng[a_i][j]$ is the rank (in the lexicographical ordering by (i, j)) of $key[a_i][j]$ in the group $G(g_l(key[a_i][j]), l)$, and*
3. *if $t = rank[a_i][j]$ and $rig[a_i][j] > 0$, then $lrm[a_i][j]$ is the number of pairs (i', j') in $G(g_{l+1}(key[a_i][j], l + 1))$ with $rig[a_{i'}][j'] < rig[a_i][j]$ and $b_l(key[a_{i'}][j']) = 0$, and*
4. *$\{rank[a_i][j] : (i, j) \in POS\} = \{0, \ldots, n - 1\}$, and*
5. *for each two pairs (i, j) and (i', j'), such that $0 \leq i, i' < p$, and $0 \leq j < s_i$, $0 \leq j' < s_{i'}$, either:*
   - $bg[i][j] = bg[i'][j']$ *and $gs[i][j] = gs[i'][j']$, or*
   - $bg[i][j] + gs[i][j] \leq bg[i'][j']$, *or*
   - $bg[i'][j'] + gs[i'][j'] \leq bg[i][j]$.

Proof. We prove Lemma 2 by induction on time slots of REGROUPING PHASE. (I.e. we show that the conditions of the lemma hold for $slot(m - 1, 0)$ and that if they hold for $slot(l, t)$ then they also hold for $next(l, t)$.) For $slot(m - 1, 0)$, the conditions of Lemma 2 are enforced by the Init procedure. Let us assume that the conditions hold for $slot(l, t)$, where $m - 1 \geq l \geq 0$ and $0 \leq t < n$. By condition 4, there is exactly one pair (a_{snd}, j') such that $rank[a_{snd}] = t$. By condition 1, (snd, j) is classified on level $l + 1$, since $t = rank[a_{snd}][j'] = bg[a_{snd}][j'] + rig[a_{snd}][j'] < bg[a_{snd}][j'] + gs[a_{snd}][j']$. Let $G' = G(g_{l+1}(key[a_{snd}][j']), l + 1)$. By conditions 5 and 1, all the pairs (i, j) in G' are classified on level $l + 1$. Let $G'_0 = \{k \in G' :$

$b_l(k) = 0\}$ and $G'_1 = G' \setminus G'_0$ (the two groups on level l that are halves of G'). The value x computed by a_{snd} is the number of pairs (i, j) in G'_0 with $rig[a_i][j] \leq rig[a_{snd}][j']$, since either $rig[a_{snd}][j'] = 0$ and a_{snd} executed $lrm[a_{snd}] \leftarrow 0$, or it follows from condition 3.

We look at variables at the beginning of $slot(l, t)$ and define three sets:

$$A = \{(i, j) : bg[a_i][j] \leq t < bg[a_i] + gs[a_i] - 1\}$$
$$B = \{(i, j) : t = bg[a_i] + gs[a_i] - 1\}$$
$$C = \{(i, j) : t < bg[a_i] \lor bg[a_i][j] + gs \leq t\}$$

Note that, by conditions 5 and 1: A, B, C is a partition of POS, and $A \cup B$ is the set of pairs that are in G', and either $A = \emptyset$ or $B = \emptyset$.

Consider the case $A \neq \emptyset$ (CASE A). Then there is exactly one pair (rcv, j) in G' such that $rank[a_{rcv}][j] = t + 1$. If $rcv \neq snd$, then $|S(a_{snd}, t)| < gs[a_{snd}][j']$ and either $j' = s_{snd} - 1$ or $bg[a_{snd}][j'+1] \neq bg[a_{snd}][j']+1$. (Otherwise $rcv = snd$ since $key[a_{snd}]$ is sorted and the keys in $key[a_{snd}]$ from G' are blocked together and have consecutive ranks.) Hence a_{snd} broadcasts $\langle x \rangle$, if necessary. (rcv, j) is preceded by $rig[a_{rcv}][j]$ pairs (i', j') in G' and x of them are in G'_0. Thus (rcv, j) should be ranked in its group on level l on position x, if $b_l(key[a_rcv][j]) = 0$, and on position $rig[a_{rcv}][j] - x$, otherwise. It follows that $rng[a_{rcv}][j]$ is updated so that condition 2 is satisfied in $next(l, t)$. The execution of $lrm[a_{rcv}] \leftarrow x$ makes the condition 3 satisfied in $next(l, t)$. Condition 1 remains satisfied in $next(l, t)$, since (in CASE A) $next(t, l) = slot(l, t + 1)$ and, for each pair (i, j) in G', $t + 1 < bg[a_i][j + gs[a_i][j]$ and (i, j) remains classified on level $l + 1$. For all pairs in C condition 1 does not change. Conditions 4 and 5 remain satisfied, since none of the involved variables is changed.

Consider the case $B \neq \emptyset$ (CASE B). For all pairs (rcv, j) in G', the station a_{rcv} has the same values bg and gs. Hence, all of them execute code for CASE B. If there is some pair (rcv, j) in G', such that $rcv \neq snd$, then $|S(a_{snd}, t)| < gs[a_{snd}][j']$. Since, in CASE B, $t = rank[a_{snd}][j'] = bg[a_{snd}][j'] + gs[a_{snd}][j'] - 1$, it follows that $rank[a_{snd}][j'] - bg[a_{snd}][j'] = rig[a_{snd}][j'] = gs[a_{snd}][j'] - 1$, and a_{snd} broadcasts $\langle x \rangle$, if necessary. Since (snd, j) is the last pair in G', the value x is the number of pairs in G'_0. Hence, each pair $(rcv, j) \in G'_0$ properly updates $gs[a_{rcv}][j]$ to x ($bg[a_{rcv}][j]$ remains unchanged), and each pair $(rcv, j) \in G'_1$ properly decreases $gs[a_{rcv}][j]$ and increases $bg[a_{rcv}][j]$ by x, for classification on level l. Besides (by condition 2) each pair (rcv, j) in G' properly updates the values of $rig[a_{rcv}][j]$ and $rank[a_{rcv}][j]$. Thus in $next(l, t)$, all the pairs in G' are classified on level l and all the pairs in C are classified as before and condition 1 holds in $next(t, l)$. Note that all ranks used by the pairs in G' in classification on level $l + 1$ are "recycled" by them in classification on level l, thus condition 4 holds in $next(l, t)$. Condition 5 holds in $next(l, t)$ since all pairs that are in the same group on level l are classified on the same level (either l or $l + 1$). Let (i, j) be the pair with $rank[a_i][j] = (t+1) \bmod n$. Condition 3 holds in $next(l, t)$ since $rig[a_i][j] = 0$. Condition 2 holds in $next(l, t)$ since (by Lemma 1(1)) also $rng[a_i][j] = 0$, and either $(t+1) = 0$ or positions with $rank < t$ had proper values of rng by induction hypothesis and no variable rng is modified in CASE B. $\square$

By Lemma 2(1), after the REGROUPING PHASE (i.e. before $slot(-1,0)$), all pairs are classified on level 0, which means the stable ranking of the keys.

4   Sorting

After the ranks of the keys have been computed we may send each key with rank r to its destination station $a_{dest(r)}$. The function $dest$ should be globally known, however its definition may depend on further applications. For example we may define $dest(r) = \lfloor p \cdot r/n \rfloor$, or $dest(r) = r \bmod p$. Procedure Route-by-ranks performs this task.

> **procedure** Route-by-ranks($\langle a_0, \ldots, a_{p-1} \rangle$)
> **begin**
> > **for** $i \leftarrow 0$ **to** $n-1$ **do**
> > > the (unique) station a_{snd} containing (unique) j such that $i = rank[a_{snd}][j]$
> > > sends message $\langle x \rangle$, where $x = key[a_{snd}][j]$ (if $a_{dest(i)} \neq a_{snd}$);
> > > the station $a_{dest(i)}$ listens (if $a_{dest(i)} \neq a_{snd}$) and stores x;
>
> **end**

Algorithm 3. Procedure Route-by-ranks

> **procedure** Counting-sort($\langle a_0, \ldots, a_{p-1} \rangle$,$m$)
> **begin**
> > Counting-rank($\langle a_0, \ldots, a_{p-1} \rangle$,$m$)
> > Route-by-ranks($\langle a_0, \ldots, a_{p-1} \rangle$)
>
> **end**

Algorithm 4. Procedure Counting-sort

5   Routing

In the case of routing, we assume that each key is a number of the station that is destination of the packet containing this key in the address field of its header. Hence we should route the packets by the keys rather than by the ranks of the keys. However, we use Counting-rank in the preprocessing phase of routing. Besides the ranks $rank[a_i][j]$, we also use the values $rig[a_i][j]$, $gs[a_i][j]$ and $n[a_i]$ computed by Counting-rank. Thus each key is from the set $\{0, \ldots, p-1\}$ and the parameter m (number of bits) is $\lceil \log_2 p \rceil$. After computing the ranks of the keys, the stations perform procedure Compute-intervals. Each station learns time interval in which it should receive its incoming packets in the procedure Finish-routing. The interval for a_i will be stored in variables $i_1[a_i]$ and $i_2[a_i]$. The packets are then broadcast in the sequence of their ranks. The whole routing is performed by the procedure Route-packets. Note that each packet is transmitted only once (in Finish-routing), while in preprocessing phase the stations transmit only the integers from the range $[0, n]$, which are usually much shorter than the whole packets.

procedure Compute-intervals($\langle a_0, \dots, a_{p-1} \rangle$)
begin
   for $i \leftarrow 0$ **to** $p-1$ **do**
      in time slot $2 \cdot i$: **begin**
         the (at most one) station a_{snd} containing (unique) j with
         $key[a_{snd}][j] = i$ and $first[a_{snd}][j] = true$ sends $\langle x \rangle$, where
         $x = rank[a_{snd}][j]$;
         a_i listens and does: **if** there was a message **then** $i_1[a_i] \leftarrow x$;
         else $i_1[a_i] \leftarrow i_2[a_i] \leftarrow (-1)$;
      end
      in time slot $2 \cdot i + 1$: **begin**
         the (at most one) station a_{snd} containing (unique) j with
         $key[a_{snd}][j] = i$ and $last[a_{snd}][j] = true$ sends $\langle x \rangle$, where
         $x = rank[a_{snd}][j]$;
         if $i_1[a_i] \neq (-1)$ **then** a_i listens and does: $i_2[a_i] \leftarrow x$;
      end
end

Algorithm 5. Procedure Compute-intervals

procedure Finish-routing($\langle a_0, \dots, a_{p-1} \rangle$)
begin
   for $i \leftarrow 0$ **to** n **do**
      in time slot i: **begin**
         the (unique) station a_{snd} containing (unique) j with $rank[a_{snd}][j] = i$
         sends packet addressed by $key[a_{snd}][j]$;
         the (unique) a_{rcv} with $i_1[a_{rcv}] \leq i \leq i_2[a_{rcv}]$ receives this packet;
         (* should be: $rcv = key[a_{snd}][j]$ *)
      end
end

Algorithm 6. Procedure Finish-routing

procedure Route-packets($\langle a_0, \dots, a_{p-1} \rangle$)
begin
   (* In each station a_i there are s_i outgoing packets sorted by their destination
   addresses, which are stored in the table $key[a_i]$ *)
   Counting-rank($\langle a_0, \dots, a_{p-1} \rangle, \lceil \log_2 p \rceil$);
   Compute-intervals($\langle a_0, \dots, a_{p-1} \rangle$);
   Finish-routing($\langle a_0, \dots, a_{p-1} \rangle$);
end

Algorithm 7. Procedure Route-packets

6 Complexities of the Procedures

Recall, that for each station a_i, s_i denotes the number of keys initially stored
by a_i, r_i is the number of distinct values of these keys, d_i is the number of keys

procedure Expand-ranks($\langle a_0, \ldots, a_{p-1} \rangle$)
begin
   Each a_i, for each $0 \leq j < s_i$, does: $first[a_i][j] \leftarrow last[a_i][j] \leftarrow false$;
   Each a_i, for each $0 \leq j' < r_i$, does: **begin**
     $first[a_i][\min P(a_i, key'[a_i][j'])] \leftarrow first'[a_i][j']$;
     $last[a_i][\max P(a_i, key'[a_i][j'])] \leftarrow last'[a_i][j']$;
   end
   Each a_i does: $lrm[a_i] \leftarrow 0$;
   for $t \leftarrow 0$ **to** $r - 2$ **do**
     the (unique) a_{snd} with table $rank'[a_{snd}]$ containing t, does: **begin**
      Let j be such that $rank'[a_{snd}][j] = t$;
      a_{snd} assigns sequential ranks $lrm[a_{snd}], \ldots, lrm[a_{snd}] + c_{snd,j} - 1$ to
      the positions $P(a_{snd}, key'[a_{snd}][j])$ of the table $rank[a_{snd}]$, where
      $c_{snd,j} = |P(a_{snd}, key'[a_{snd}][j])|$;
      a_{snd} sends message $\langle x \rangle$, where $x = lrm[a_{snd}] + c_{snd,j}$;
     end
     the (unique) a_{rcv} with table $rank'[a_{rcv}]$ containing $t + 1$, receives $\langle x \rangle$ and
     does: $lrm[a_{rcv}] \leftarrow x$;
   the (unique) a_{snd} with table $rank'[a_{snd}]$ containing $r - 1$, does: **begin**
     Let j be such that $rank'[a_{snd}][j] = r - 1$;
     a_{snd} assigns sequential ranks $lrm[a_{snd}], \ldots, lrm[a_{snd}] + c_{snd,j} - 1$ to the
     positions $P(a_{snd}, key'[a_{snd}][j])$ of the table $rank[a_{snd}]$, where
     $c_{snd,j} = |P(a_{snd}, key'[a_{snd}][j])|$;
     a_{snd} sends message $\langle x \rangle$, where $x = lrm[a_{snd}] + c_{snd,j}$.
   end
   each $a_i \neq a_{snd}$ listens to $\langle x \rangle$;
   each a_i does: $n[a_i] \leftarrow x$;
end

Algorithm 8. Procedure Expand-ranks

for which a_i is destination, and q_i is the number of stations that initially stored such keys. Also p is the number of stations, $n = \sum_{i=0}^{p-1} s_i$ and $r = \sum_{i=1}^{p-1} r_i$.

Lemma 3. *For* Init *the energetic cost of listening, for each a_i, is at most 2 and the energetic cost of sending is at most 1. Time of* Init *is p.*

Lemma 4. *For* Counting-rank, *for each a_i, the energetic cost of listening is at most $2m \cdot r_i + 2$ and the energetic cost of sending is at most $m \cdot r_i + 1$. Time of* Counting-rank *is $m \cdot n + p$.*

Proof. Ranks of the keys from the same group g are continuous in a single station a_i, and all keys with the same value v are always in the same group. For each such group g, a_i has to listen only to the predecessor of its key with the lowest rank in g and to the last element in g, if it is in another station. Similar arguments can be used to estimate the cost of sending. Time complexity is easily seen from the code of the procedure. □

Lemma 5. *For* Route-by-ranks, *for each a_i, the energetic cost of listening is at most d_i and the cost of sending is s_i. The time of* Route-by-ranks *is n.*

Lemma 6. *For* Counting-sort, *for each* a_i, *the energetic cost of listening is* $2m \cdot r_i + d_i + 2$ *and the cost of sending is* $m \cdot r_i + s_i + 1$. *(The total energetic cost is:* $3m \cdot r_i + d_i + s_i + 3$.*) The time is* $m \cdot n + n + p$.

Lemma 7. *For* Compute-intervals, *for each* a_i, *the energetic cost of listening is* 2 *and energetic cost of sending is at most* $2r_i$. *The time is* $2p$.

Lemma 8. *For* Finish-routing, *for each* a_i, *the energetic cost of listening is* d_i, *and the energetic cost of sending is* s_i. *The time is* n.

Lemma 9. *For* Route-packets, *for each* a_i, *the energetic cost of listening is* $2m \cdot r_i + d_i + 4$, *and the cost of sending is* $m \cdot r_i + s_i + 2r_i + 1$, *where* $m = \lceil \log_2 p \rceil$ *and* $r_i \leq p - 1$ *(no station sends packets to itself). (Total energetic cost is:* $3m \cdot r_i + s_i + d_i + 2r_i + 5$*). The time is* $m \cdot n + n + 3p$.

7   Accelerating Counting-Rank

Time complexity of Counting-rank contains component $m \cdot n$. Note that, in the case of routing, $r \leq p(p - 1)$ (each station sends packets to at most $p - 1$ other stations) and $m = \lceil \log_2 p \rceil$. We may expect that n is much larger than r. In this section we show how to replace this component with $(m + 1) \cdot r$ while the energetic cost for each a_i is increased by at most $2r_i + 1$. By Compressed-counting-rank we denote the procedure Counting-rank with the code modified as follows: Each station a_i pretends that it contains only one key of given value (i.e. it uses key', $rank'$, $last'$, $first'$, r_i and r instead of key, $rank$, $last$, $first$, s_i and n, respectively.) The code of the sub-procedure Init is modified the same way. The computed results are stored in $rank'$, $first'$ and $last'$. At the end of Compressed-counting-rank we add procedure Expand-ranks (Algorithm 8) which computes ranks and proper values $first$ and $last$, for all keys, and proper value of n in each station. Let $P(a_i, k) = \{j \mid key[a_i][j] = k\}$. The time of Expand-ranks is r and its energetic cost for each a_i is $2r_i + 1$. (Each a_i may need to listen and send at most once for each its key value and listen to the last message.) Let us call the resulting algorithm Accelerated-Routing.

Lemma 10. *For* Accelerated-Routing, *for each* a_i, *the energetic cost of listening and sending is* $3\lceil \log_2 p \rceil \cdot r_i + s_i + d_i + 4r_i + 6$, *where* $r_i \leq p - 1$. *The time is* $\lceil \log_2 p \rceil \cdot r + n + r + 3p$.

References

1. Datta, A., Zomaya, A.Y.: An Energy-Efficient Permutation Routing Protocol for Single-Hop Radio Networks. IEEE Trans. Parallel Distrib. Syst. 15, 331–338 (2004)
2. Kik, M.: Merging and Merge-sort in a Single Hop Radio Network. In: Wiedermann, J., Tel, G., Pokorný, J., Bieliková, M., Štuller, J. (eds.) SOFSEM 2006. LNCS, vol. 3831, pp. 341–349. Springer, Heidelberg (2006)

3. Kik, M.: Sorting Long Sequences in a Single Hop Radio Network. In: Královič, R., Urzyczyn, P. (eds.) MFCS 2006. LNCS, vol. 4162, pp. 573–583. Springer, Heidelberg (2006)
4. Nakano, K.: An Optimal Randomized Ranking Algorithm on the k-channel Broadcast Communication Model. In: ICPP 2002, pp. 493–500 (2002)
5. Nakano, K., Olariu, S., Zomaya, A.Y.: Energy-Efficient Permutation Routing in Radio Networks. IEEE Transactions on Parallel and Distributed Systems 12, 544–557 (2001)
6. Nakano, K., Olariu, S., Zomaya, A.Y.: Energy-Efficient Routing in the Broadcast Communication Model. IEEE Trans. Parallel Distrib. Syst. 13, 1201–1210 (2002)
7. Singh, M., Prasanna, V.K.: Optimal Energy Balanced Algorithm for Selection in Single Hop Sensor Network. In: IEEE International Workshop on Sensor Network Protocols and Applications SNPA ICC (May 2003)
8. Singh, M., Prasanna, V.K.: Energy-Optimal and Energy-Balanced Sorting in a Single-Hop Sensor Network. In: IEEE Conference on Pervasive Computing and Communications PERCOM (March 2003)
9. Compendium of Large-Scale Optimization Problems (DELIS, Subproject 3), http://ru1.cti.gr/delis-sp3/

Intrusion Detection of Sinkhole Attacks
in Wireless Sensor Networks

Ioannis Krontiris, Tassos Dimitriou, Thanassis Giannetsos,
and Marios Mpasoukos

Athens Information Technology,
P.O.Box 68, 19.5 km Markopoulo Ave.,
GR- 19002, Peania, Athens, Greece
{ikro,tdim,agia,mamp}@ait.edu.gr

Abstract. In this paper, we present an Intrusion Detection System designed for wireless sensor networks and show how it can be configured to detect Sinkhole attacks. A Sinkhole attack forms a serious threat to sensor networks. We study in depth this attack by presenting how it can be launched in realistic networks that use the MintRoute protocol of TinyOS. MintRoute is the most widely used routing protocol in sensor network deployments, using the link quality metric to build the corresponding routing tree. Having implemented this attack in TinyOS, we embed the appropriate rules in our IDS system that will enable it to detect successfully the intruder node. We demonstrate this in our own sensor network deployment and we also present simulation results to confirm the effectiveness and accuracy of the algorithm in the general case of random topologies.

1 Introduction

Most of the applications in wireless sensor networks (WSN) require the unattended operation of a large number of sensors. This fact along with the limited computational and communication resources of their nodes make them susceptible to attacks. Sensor networks cannot rely on human intervention to face an adversary's attempt to compromise the network or hinder its proper operation. Instead, an autonomic response of the network that relies on the embedded pre-programmed policies and a coordinated, cooperative behavior is the most effective way to gain maximum advantage against adversaries.

So far, research in sensor networks security has made certain progress in providing specialized security mechanisms, like key establishment [1], secure localization [2], or secure aggregation [3]. Also, security protocols have been designed with the goal of protecting a sensor network against particular attacks, like selective forwarding [4], sinkhole [5] or wormhole [6] attacks. However, all these protocols fall prey to *insider* attacks, in which the attacker has compromised and retrieved the cryptographic material of a number of nodes. Because of their resource constraints, sensor nodes usually cannot deal with such strong adversaries. So what is needed is a second line of defense: An *Intrusion Detection*

M. Kutyłowski et al. (Eds.): ALGOSENSORS 2007, LNCS 4837, pp. 150–161, 2008.

System (IDS) that can *detect* a third party's attempts of exploiting protocol weaknesses and *warn* of malicious behavior. Using an IDS, the network will be able to respond and isolate the intruder in order to protect and guarantee its normal operation.

In [7], we proposed an IDS for sensor networks which is designed to work with only partial and localized information available in each node. In particular, we concentrated on how such an IDS could detect *blackhole* and *selective forwarding* attacks. The nodes simply monitor their neighborhood and collaborate with each other sharing valuable information that eventually leads to the successful detection of the attack.

In this paper we extend this IDS system so that it can detect *sinkhole* attacks, a particularly severe attack that prevents the base station from obtaining complete and correct sensing data, thus forming a serious threat to higher-layer applications. By showing how the detection of such attacks can be integrated in the IDS, we move a step further towards a complete intrusion detection solution for sensor networks.

Current routing protocols in sensor networks are susceptible to sinkhole attacks [8]. This is because these protocols were not designed having security threats in mind. In this paper, we concentrate on MintRoute, which is among the most widely used routing protocols in TinyOS. MintRoute is used in most real sensor networks deployments today, as for example in [9,10,11], therefore it is important to guarantee protection of such networks from sinkhole attacks. To emphasize this further, we also demonstrate how easily it is for an intruder to launch a sinkhole attack against a network having MintRoute as its underlying routing protocol. We implemented both the attack and the IDS in TinyOS to demonstrate the effectiveness and accuracy of the intrusion detection process.

The remainder of this paper is organized as follows. Section 2 references related work. In Section 3, we review MintRoute emphasizing on its basic characteristics that an attacker could exploit to launch a sinkhole attack. In Section 4, we present in detail how this attack can be realized by an intruder node, and in Section 5, we present the architecture of our IDS system. Finally, in Section 6, we present simulation results that show the behavior of the IDS in several random topologies and our demonstration on a realistic sensor network deployment.

2 Related Work

There are currently only a few studies in the area of intrusion detection in wireless sensor networks. Da Silva *et al.* [12] and Onat and Miri [13] propose two similar IDS systems, where certain monitoring nodes in the network are functioning as watchdogs for their neighbors, looking for intruders. They listen to messages in their radio range and store in a buffer specific message fields that might be useful to an IDS system running within a sensor node, but no details are given how this system works. In these architectures, there is no collaboration among the monitor nodes. It is concluded from both papers that the buffer size is an important factor that greatly affects the rate of false alarms.

A first approach on the intrusion detection of sinkhole attacks has been presented in [5]. However this approach involves the base station in the detection process, resulting in a high communication cost for the protocol. Furthermore, it cannot be generalized to an IDS that could detect more types of attacks, as it is designed just for that particular attack. We believe that having a more general IDS architecture first can be enriched later with the support for more attacks leading to a complete solution.

3   Mintroute

MintRoute is one of the most commonly used routing protocol in TinyOS. It uses link quality estimates as the routing cost metric to build the routing tree towards the base station. For the calculation of these link estimates, each node periodically transmits a packet, called *"route update"*.

Each node estimates the link quality of its neighbors based on the *packet loss* of the route update packets received from each corresponding neighbor. The list of these estimates for each neighbor is broadcasted by the node periodically in its own route update packets.

Every node maintains a Neighbor Table and updates it when it receives a route update packet. This table stores a list with the IDs of all neighboring nodes and their corresponding link costs. The node chooses its "parent node" to be the one with the best link quality in the Neighbor Table. Note that the hop distance of each neighbor to the base station is not taken under consideration in choosing the parent.

The parent changing mechanism is triggered each time the link quality of one or more nodes becomes 75% better than the link quality of the current parent, or the link quality of the current parent drops below 25 (in absolute value). In such case, the node with the highest quality becomes the new parent. However, if two of such candidate nodes happens to have the same link quality, the new parent will be the first one found in the Neighbor Table.

4   Sinkhole Attack

In a Sinkhole attack [8] a compromised node tries to draw all or as much as possible traffic from a particular area, by making itself look attractive to the surrounding nodes with respect to the routing metric. As a result, the adversary manages to attract all traffic that is destined to the base station. By taking part in the routing process, she can then launch more severe attacks, like selectively forwarding, modifying or even dropping the packets coming through.

A compromised node does not necessarily have to target other nodes from areas outside its neighborhood in order to control traffic. The adversary needs only to launch the sinkhole attack from a node as close as possible to the base station. In this case, by having the neighboring nodes choose the intruder as their parent, all the traffic coming from their descendants will also end up in the sinkhole. So the attack can be very effective even if it is launched locally, with small effort from the side of the attacker.

In the case of a routing protocol, like MintRoute, that uses link estimates as the routing metric, the compromised node launching the sinkhole attack will try to persuade its neighbors to change their current parents and choose the sinkhole node as their new one, by trying to make these parents look like they have much worse link quality than itself. Note that in this case of MintRoute, the attacker cannot launch a sinkhole attack by advertising that is has a lower hop count to the base station, as this metric is not used in the routing protocol. So the attacker needs to come up with more sophisticated ways.

Moreover, by just advertising a high link quality to the other nodes may not be enough to make them change their parents, since most of these routing protocols try to be robust, meaning that they don't allow the nodes change parents frequently and for no good reason. For example, when a node changes its parent, that could create a routing cycle in the network, which is followed by an extra cost to resolve it. Therefore, aside from advertising a high link quality for itself, the attacking node needs to make the current parents look like they have a very poor link quality, which will trigger the parent changing mechanism in their children. Then the new parent to be chosen will be the sinkhole node.

One sinkhole attack using the MintRoute as the underlying routing protocol is presented in [14]. The method of this attack is to change the link quality estimates sent by the nodes, within the route update packets. To do that, the attacker listens to the route update messages from its neighbors, alters them and replays them impersonating the original sender. Even if there is an underlying key mechanism which nodes can use to communicate securely with each other, most probably the attacker will be using a broadcast key shared with the nodes to be able to overhear change and send these packets.

Let's take for example the case shown in Figure 1, where node C is the attacker and node B is the current parent of node A. Node C has sent its own route update packets advertising a fake link quality (at the maximum value of 255), but this is not enough to make node A change its parent. Therefore, when it receives the route update packet of node A, it changes the link quality of node

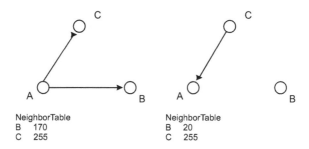

Fig. 1. The two phases of sinkhole attack. In the first phase node C (attacker) receives the route update packet of node A and in the second phase it sends the forged packet to A impersonating B.

B to a low value and sends it back to A as a unicast packet, impersonating B. Upon receiving this packet, node A thinks it is a route update packet from B and updates the corresponding entry in the Neighbor Table. This will trigger the parent changing mechanism and since the link quality of node B is below 25, that node will be ignored and node C will be chosen.

After performing the above attack for all of its neighbors, eventually the Sinkhole node will attract most (if not all) of the networks traffic.

5   Intrusion Detection

In this section we propose an IDS for sensor networks that is able to detect an ongoing sinkhole attack. For the design of our solution we have assumed a routing layer that is based on link quality metrics to form a routing tree towards the base station.

The intrusion detection system follows a distributed architecture. It is composed of identical IDS clients running in each node in the network. Then the IDS clients communicate with each other in order to reach a conclusion on an intrusion event. The functionality of each IDS client can be summarized as following:

- *Network Monitoring*: Each IDS client listens on the network and captures and examines individual packets passing from its immediate neighborhood in real time. Since all communication in a WSN is conducted over the air, and each node can overhear the traffic in its neighborhood, this is a natural audit source for the IDS client.
- *Intrusion Detection*: Each IDS client follows a specification-based approach in order to detect attacks, i.e., it detects deviations from normal behavior based on user defined rules. The network administrator have to define and embed in the motes the corresponding rules for each attack that the IDS should detect. In this paper we define the rules for the sinkhole attack, which we will present shortly.
- *Decision Making*: Due to its myopic vision around its neighborhood, a node may not be able to make a final decision whether a node is indeed an intruder. But even if it is, it cannot be trusted by the network, as it can be malicious itself. Therefore, if an anomaly is detected by an IDS client then a cooperative mechanism is initiated with the neighboring nodes so that all of them come to a mutual conclusion.
- *Action*: Every node has a response mechanism that allow it to respond to an intrusion situation.

Based on these functions we build the architecture of the IDS client around five conceptual modules, as shown in Figure 2. Each module is responsible for a specific function, which we describe in the sections below. The IDS clients are identical in each node and they can exchange messages with clients in neighboring nodes.

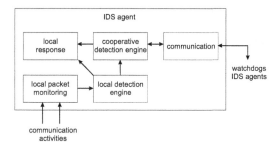

Fig. 2. The building blocks of the IDS client existing in each sensor node

5.1   Local Packet Monitoring

This module gathers audit data to be provided to the local detection module. Audit data in a sensor network IDS system can be the communication activities within its radio range. This data can be collected by listening *promiscuously* to neighboring nodes' transmissions. As sensor nodes have this capability, this can be very useful for intrusion detection.

In particular, in our IDS design we require that for each node in the network, any of its neighbors listening to the packets that this node is sending or receiving will participate in the intrusion detection procedure. Therefore, the neighbors of a node function as watchdogs for that node. As we will see in Section 5.3, one watchdog is *not enough* to detect a sinkhole attack, but if all neighbors contribute their point of view to the rest of them, then the picture becomes complete, and the attacker is revealed.

More importantly, there is no need for the watchdogs to store the overheard packets or any other information in their memory. It is just enough to temporarily buffer each packet in order to apply the rules defined by the local detection engine and see if any of these rules are satisfied. Then the packet can be discarded. No historical or statistical data need to be kept in the node's memory.

5.2   Local Detection Engine

The local detection engine stores and applies all the specifications that describe what is a correct operation and monitors audit data with respect to these constraints, in order to identify any deviations from normal behavior. These specifications are defined in the form of rules, specified by the developer, since we want to avoid the overhead of training the network to what is a normal behavior.

In order to detect the sinkhole attack we add two rules that will trigger an alert whenever the malicious node tries to impersonate another node, according to the attack we described in Section 4. The intuition is that route update packets should originate only from their legitimate sender and the nodes should defend against impersonation attacks.

Rule 1: *"For each overhead route update packet check the sender field, which must be different than your node ID. If this is not the case, produce an alert and broadcast it to your neighbors."*

Rule 2: *"For each overhead route update packet check the sender field, which must be the node ID of one of your neighbors. If this is not the case, produce an alert and broadcast it to your neighbors."*

For a node that detects an anomaly according to the above rules it is only an indication that a sinkhole attack is in process. However there is no way to know which node is trying to launch the attack, since the sender field is altered. The only conclusion that can be drawn so far is that the attacker is one of the neighboring nodes, since the route update packets are only broadcasted locally. So, we need to rely on the cooperation of the nodes to reduce the candidates to one node, i.e. the attacker.

5.3 Cooperative Detection Engine

The problem we need to solve is how the intruder node will be revoked from the network. Basing this decision on an individual node is not sufficient for two reasons:

1. The node who makes the final decision can be compromised itself. Then it could choose not to revoke an attacking node or revoke a legitimate one. So, the decision should be collaborative, and should come from all the nodes that are involved, i.e. the watchdogs.
2. In the case of sinkhole attack there is not enough information in only one node to conclude on the attacker.

Therefore, we need a cooperative detection engine that will guide the nodes through a safe conclusion on which node is the intruder so that it can be revoked by the network. In particular, we are going to exploit the fact that when a malicious node launches a sinkhole attack, one of the two rules in the local detection engine will be triggered at several of its neighbors. If these neighbors collaborate, it turns out that they can identify the attacker.

The collaborative approach consists of having each watchdog node broadcasting its neighbors list. As we said, each watchdog that produces a local alert can conclude nothing more than that the attacker is one of its neighbors. However, if all the nodes that produce the alert communicate their neighbors to each other, the attacker has to be one of the nodes in the *intersection* of these sets.

So, to set it more formally, if there is evidence of intrusion produced at the local detection engine, the cooperative detection module broadcasts an alert to the neighboring nodes. The alert is composed by the list of the IDs of the sender's *neighbors*. Upon receiving such an alert, and provided that it is a watchdog itself, a node excludes from the potential attackers all the node IDs in the alert that are not part of its own neighbor list. Thus it performs an intersection between its own neighbor list and the node list found in the alert. The result will be

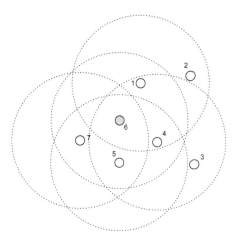

Fig. 3. An example topology where node 6 is the attacker. If each node broadcasts an alert with the list of its neighbors, the intersection of the alerts each node receives is node 6.

stored and used for the intersection with the next alert that the node is going to receive. Thus each time a watchdog is receiving an alert, the intersection will give an even smaller set of nodes. If at the end there is only one node left at the result, that node is the attacker.

The intuition in this is that each time a node broadcast an alert with its neighbors, this set is a set that includes the attacker. By exchanging these sets and performing the intersections, nodes are looking to find which nodes are common within the sets. If some nodes manage to reduce this set down to one node, then they can be sure about the intruder's identity. Let's see this in an example.

In Figure 3, the attacker is node 6. Each node broadcasts an alert including its neighbor list. Then, it intersects the lists from the alerts it receives along with its own neighbor list. So, we will have the following results:

Node 1: $\{2,4,6\} \cap \{1,3,5,6\} = \{6\}$
Node 4: $\{1,3,5,6\} \cap \{2,4,6\} \cap \{4,6,7\} = \{6\}$.
Node 5: $\{4,6,7\} \cap \{1,3,5,6\} \cap \{5,6\} = \{6\}$.
Node 7: $\{5,6\} \cap \{4,6,7\} = \{6\}$

Each node remains with a set of one node, the attacker. It could be the case (not demonstrated in the example) that the set had more than one node and then no conclusion could be drawn. However, we will see in the experimental section that more than 75% of the attacker's watchdogs will manage to successfully detect it, meaning that they will end up with only one node ID as the result of the cooperative detection process.

5.4 Local Response

Once the watchdogs are aware that an intrusion has taken place and have detected the compromised node, appropriate actions are taken by the local response module. The first action is to cut off the intruder as much as possible and isolate the compromised node. After that, proper operation of the network must be restored. This may include changes in the routing paths, updates of the cryptographic material (keys, etc.) or restoring part of the system using redundant information distributed in other parts of the network. Autonomic behavior of sensor networks means that these functions must be performed without human intervention and within finite time.

Depending on the confidence and the type of the attack, we categorize the response to two types:

- *Direct response*: Excluding the suspect node from any paths and forcing regeneration of new cryptographic keys with the rest of the neighbors.
- *Indirect response*: Notifying the base station about the intruder or reducing the quality estimation for the link to that node, so that it will gradually loose its path reliability.

We will not go in more depth regarding the local response process, since for this paper we are more concentrated on the former steps of the intrusion detection. However, let us note that in any case, the local response cannot be based on the claims of one of a few nodes, since the attacker may have compromised and use more than one node during the attack. Therefore, we require that the majority of the attacker's neighbors have successfully detect the sinkhole node. Otherwise we count the case as a false negative, i.e. the network could not reach a safe conclusion. We measure the false negative rate in the experimental section that follows.

6 Experimental Evaluation

We have simulated a sensor network of 100 nodes placed uniformly at random in order to test our proposed intrusion detection system. For each run of the simulation, we chose at random one node to launch a sinkhole attack. This way we could have the watchdogs of that node apply the intrusion detection and monitor its behavior.

To measure the success of the IDS system on identifying the intruder, we first run the simulation 1000 times and produced the average number of the watchdogs that ended up with only one ID (the intruder ID) as the result of the alerts intersection. As we see in Figure 4(a), the majority of the watchdogs were able to detect the malicious node. Let's also note that as the network density increases, the results improve, meaning that bigger portion of the watchdogs manage to identify the attacker. This is due to the fact that more watchdogs are within the range of each other, so the intersection of the alerts is more likely that it will produce a unique node ID. For example, for a network density of

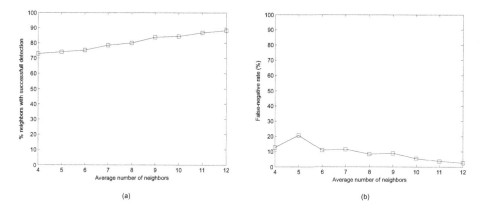

(a)   (b)

Fig. 4. (a) The percentage of watchdogs that successfully detect the attacker for different network densities. (b) False-negative rate for different network densities.

6 neighbors on average, 75% of them will identify the attacker, while for 12 neighbors, the percentage goes up to 88.3%.

Next we measured the false negative rate of our IDS system. We define a false negative as the case where the majority of the watchdogs were not able to conclude to one node ID and therefore identify the attacker. This is possible if the topology is such that the intersection of the alerts that each watchdog receives produces a set of more than one node. However, as we see in Figure 4(b), the probability that less than half of the watchdogs remain inconclusive is very low and becomes even lower as the network density increases. For example, for a network density of 6 neighbors on average, the false negative rate is 11%, while for 10 neighbors it is 5.3%.

Next we implemented our IDS system in TinyOS in order to evaluate it in a realistic deployment of sensor nodes (Mica2) and analyze its memory requirements per node. In particular, it required 1.5 KB in RAM (out of 4 KB available) and 3.9 KB in ROM (out of 128 KB available), which are realistic memory overheads for an intrusion detection system. However we are not aware of any other IDS implementation in TinyOS to use as a comparative measure.

We programmed a node (node 5) to launch the sinkhole attack as described in Section 4. Then we programmed the rest of the nodes with MintRoute so that they could form a routing tree as shown in Figure 5. We also wired our IDS client in each of these nodes to see if they could detect the attack and identify the intruder node.

Indeed, nodes 1, 4, 6 and 8 produced intrusion alerts and successfully identified node 5 as the attacker. Let's consider node 8 for example. Its neighbors are nodes $\{1, 4, 5, 6\}$. The attacker tried to impersonate it, sending routing update packets to its children, node 4, so an alert was produced due to rule 1, as we described in Section 5.2. Moreover, it received the alerts from nodes 4, 8 and 1, which

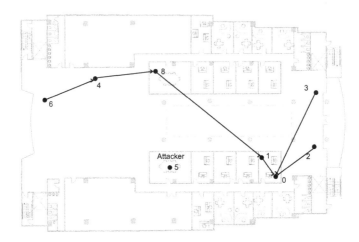

Fig. 5. The routing tree formed by the sensor nodes using MintRoute. Node 5 is the attacker launching a sinkhole attack.

also produced alerts when the attacker tried to attract the rest of the nodes. The intersection of these alerts with its own neighbors resulted in just node 5: $\{1, 4, 5, 6\} \cap \{4, 8, 5\} \cap \{8, 6, 5\} \cap \{0, 2, 3, 5, 8\} = \{5\}$.

Similarly, for the rest of the nodes we had

Node 1: $\{0, 2, 3, 5, 8\} \cap \{4, 1, 0, 5, 6\} = \{5\}$
Node 4: $\{8, 6, 5\} \cap \{4, 8, 5\} \cap \{4, 1, 0, 5, 6\} = \{5\}$
Node 6: $\{4, 8, 5\} \cap \{4, 1, 0, 5, 6\} \cap \{8, 6, 5\} = \{5\}$

Let us note that the attacker's neighbors are $\{1, 2, 4, 6, 8\}$. Therefore, 4 out of its 5 neighbors successfully detected the attack. No rules were triggered for nodes 2 and 3, so they remained out of the process.

7   Conclusions

In this paper, we described a model for a distributed intrusion detection system that uses a large number of autonomous, but localized, cooperating agents in order to detect a node launching a sinkhole attack. The nodes use coordinated surveillance by incorporating inter-agent communication and distributed computing in decision making to identify characteristic signs of the attack, and raise an appropriate alarm. In particular, we concentrated on the sinkhole attack against routing update protocols based on link quality like MintRoute and we described the appropriate specifications that need to be implemented by the IDS system so that it can detect such attacks. We believe this set of principles can be used as a valuable tool for developing more robust and secure sensor networks in the future and enable further research in the area.

References

1. Camtepe, S., Yener, B.: Key distribution mechanisms for wireless sensor networks: a survey. Technical Report 05-07, Rensselaer Polytechnic Institute, Troy, New York (March 2005)
2. Lazos, L., Poovendran, R.: Serloc: Robust localization for wireless sensor networks. ACM Transactions on Sensor Networks 1(1), 73–100 (2005)
3. Dimitriou, T., Krontiris, I.: Secure In-network Processing in Sensor Networks. In: Security in Sensor Networks, pp. 275–290. CRC Press, Boca Raton, USA (2006)
4. Yu, B., Xiao, B.: Detecting selective forwarding attacks in wireless sensor networks. In: Proceedings of the 20th International Parallel and Distributed Processing Symposium (SSN2006 workshop), Rhodes, Greece, pp. 1–8 (April 2006)
5. Ngai, E.C.H., Liu, J., Lyu, M.R.: On the intruder detection for sinkhole attack in wireless sensor networks. In: ICC 2006. Proceedings of the IEEE International Conference on Communications, Istanbul, Turkey (2006)
6. Hu, Y.C., Perrig, A., Johnson, D.B.: Packet leashes: A defense against wormhole attacks in wireless ad hoc networks. In: INFOCOM 2003. Proceedings of the Twenty-Second Annual Joint Conference of the IEEE Computer and Communications Societies, San Francisco, CA, USA (2003)
7. Krontiris, I., Dimitriou, T., Freiling, F.C.: Towards intrusion detection in wireless sensor networks. In: Proceedings of the 13th European Wireless Conference, Paris, France (April 2007)
8. Karlof, C., Wagner, D.: Secure routing in wireless sensor networks: Attacks and countermeasures. AdHoc Networks Journal 1(2–3), 293–315 (2003)
9. Baggio, A.: Wireless sensor networks in precision agriculture. In: REALWSN 2005. Proceeding of the Workshop on Real-World Wireless Sensor Networks, Stockholm, Sweden (June 2005)
10. Werner-Allen, G., Lorincz, K., Welsh, M., Marcillo, O., Johnson, J., Ruiz, M., Lees, J.: Deploying a wireless sensor network on an active volcano. IEEE Internet Computing 10(2), 18–25 (2006)
11. Schmid, T., Dubois-Ferrière, H., Vetterli, M.: SensorScope: Experiences with a Wireless Building Monitoring Sensor Network. In: REALWSN 2005. Proceeding of the Workshop on Real-World Wireless Sensor Networks, Stockholm, Sweden (June 2005)
12. da Silva, A.P., Martins, M., Rocha, B., Loureiro, A., Ruiz, L., Wong, H.C.: Decentralized intrusion detection in wireless sensor networks. In: Q2SWinet 2005. Proceedings of the 1st ACM international workshop on Quality of service & security in wireless and mobile networks, pp. 16–23. ACM Press, New York (2005)
13. Onat, I., Miri, A.: An intrusion detection system for wireless sensor networks. In: Proceeding of the IEEE International Conference on Wireless and Mobile Computing, Networking and Communications, Montreal, Canada, August 2005, vol. 3, pp. 253–259 (2005)
14. Datema, S.: A case study of wireless sensor network attacks. MSc thesis, Delft University of Technology (2005)

Author Index

Lecture Notes in Computer Science

Sublibrary 5: Computer Communication Networks and Telecommunications

Vol. 4269: R. State, S. van der Meer, D. O'Sullivan, T. Pfeifer (Eds.), Large Scale Management of Distributed Systems. XIII, 282 pages. 2006.

Vol. 4268: G. Parr, D. Malone, M. Ó Foghlú (Eds.), Autonomic Principles of IP Operations and Management. XIII, 237 pages. 2006.

Vol. 4267: A. Helmy, B. Jennings, L. Murphy, T. Pfeifer (Eds.), Autonomic Management of Mobile Multimedia Services. XIII, 257 pages. 2006.

Vol. 4240: S.E. Nikoletseas, J.D.P. Rolim (Eds.), Algorithmic Aspects of Wireless Sensor Networks. X, 217 pages. 2006.

Vol. 4238: Y.-T. Kim, M. Takano (Eds.), Management of Convergence Networks and Services. XVIII, 605 pages. 2006.

Vol. 4235: T. Erlebach (Ed.), Combinatorial and Algorithmic Aspects of Networking. VIII, 135 pages. 2006.

Vol. 4217: P. Cuenca, L. Orozco-Barbosa (Eds.), Personal Wireless Communications. XV, 532 pages. 2006.

Vol. 4195: D. Gaiti, G. Pujolle, E.S. Al-Shaer, K.L. Calvert, S. Dobson, G. Leduc, O. Martikainen (Eds.), Autonomic Networking. IX, 316 pages. 2006.

Vol. 4124: H. de Meer, J.P.G. Sterbenz (Eds.), Self-Organizing Systems. XIV, 261 pages. 2006.

Vol. 4104: T. Kunz, S.S. Ravi (Eds.), Ad-Hoc, Mobile, and Wireless Networks. XII, 474 pages. 2006.

Vol. 4074: M. Burmester, A. Yasinsac (Eds.), Secure Mobile Ad-hoc Networks and Sensors. X, 193 pages. 2006.

Vol. 4033: B. Stiller, P. Reichl, B. Tuffin (Eds.), Performability Has its Price. X, 103 pages. 2006.

Vol. 4026: P.B. Gibbons, T. Abdelzaher, J. Aspnes, R. Rao (Eds.), Distributed Computing in Sensor Systems. XIV, 566 pages. 2006.

Vol. 4003: Y. Koucheryavy, J. Harju, V.B. Iversen (Eds.), Next Generation Teletraffic and Wired/Wireless Advanced Networking. XVI, 582 pages. 2006.

Vol. 3996: A. Keller, J.-P. Martin-Flatin (Eds.), Self-Managed Networks, Systems, and Services. X, 185 pages. 2006.

Vol. 3976: F. Boavida, T. Plagemann, B. Stiller, C. Westphal, E. Monteiro (Eds.), NETWORKING 2006. Networking Technologies, Services, and Protocols; Performance of Computer and Communication Networks; Mobile and Wireless Communications Systems. XXVI, 1276 pages. 2006.

Vol. 3970: T. Braun, G. Carle, S. Fahmy, Y. Koucheryavy (Eds.), Wired/Wireless Internet Communications. XIV, 350 pages. 2006.

Vol. 3964: M.Ü. Uyar, A.Y. Duale, M.A. Fecko (Eds.), Testing of Communicating Systems. XI, 373 pages. 2006.

Vol. 3961: I. Chong, K. Kawahara (Eds.), Information Networking. XV, 998 pages. 2006.

Vol. 3912: G.J. Minden, K.L. Calvert, M. Solarski, M. Yamamoto (Eds.), Active Networks. VIII, 217 pages. 2007.

Vol. 3883: M. Cesana, L. Fratta (Eds.), Wireless Systems and Network Architectures in Next Generation Internet. IX, 281 pages. 2006.

Vol. 3868: K. Römer, H. Karl, F. Mattern (Eds.), Wireless Sensor Networks. XI, 342 pages. 2006.

Vol. 3854: I. Stavrakakis, M. Smirnov (Eds.), Autonomic Communication. XIII, 303 pages. 2006.

Vol. 3813: R. Molva, G. Tsudik, D. Westhoff (Eds.), Security and Privacy in Ad-hoc and Sensor Networks. VIII, 219 pages. 2005.

Vol. 3462: R. Boutaba, K.C. Almeroth, R. Puigjaner, S. Shen, J.P. Black (Eds.), NETWORKING 2005. XXX, 1483 pages. 2005.